AF691415

Microelectronics

Microelectronics

Editor

Tudor Volkov

scitus academics

Microelectronics
Edited by **Tudor Volkov**

Printed in 2017

ISBN: 978-1-68117-163-0
Library of Congress Control Number: 2015950368

© 2016 by
SCITUS Academics LLC,
616, Corporate Way, Suite 2, 4766,
Valley Cottage, NY 10989

www.scitusacademics.com

This book contains information obtained from highly regarded resources. Copyright for individual articles remains with the authors as indicated. All chapters are distributed under the terms of the Creative Commons Attribution License, which permits unrestricted use, distribution, and reproduction in any medium, provided the original author and source are credited.

Notice

Reasonable efforts have been made to publish reliable data and views articulated in the chapters are those of the individual contributors, and not necessarily those of the editors or publishers. Editors or publishers are not responsible for the accuracy of the information in the published chapters or consequences of their use. The publisher believes no responsibility for any damage or grievance to the persons or property arising out of the use of any materials, instructions, methods or thoughts in the book. The editors and the publisher have attempted to trace the copyright holders of all material reproduced in this publication and apologize to copyright holders if permission has not been obtained. If any copyright holder has not been acknowledged, please write to us so we may rectify.

Table of Content

Preface

Microelectronics, as the name suggests, is a subfield of electronics and relates to the study and microfabrication of very small electronic designs and components. Usually, but not always, this means micrometre-scale or smaller. These devices are frequently made from semiconductor materials. Many components of normal electronic design are available in a microelectronic equivalent. These comprise transistors, capacitors, inductors, resistors, diodes and (naturally) insulators and conductors can all be found in microelectronic devices. Unique wiring techniques such as wire bonding are also frequently used in microelectronics because of the unusually small size of the components, leads and pads.

Microelectronics is fundamentally a multi-disciplinary field of research, exploring the physics, materials and chemistry required to make devices work. It is also multidisciplinary with regard to its wide variety of applications, as it plays a vital role in all fields of innovation, ranging from advanced health care to telecommunications and smart grids. The ever-increasing demand for processing power, sensing capabilities and miniaturisation makes microelectronics a highly innovative research field.

As techniques improve, the scale of microelectronic components continues to decrease. At smaller scales, the relative impact of intrinsic circuit properties such as interconnections may become noteworthy. These are called parasitic effects, and the goal of the microelectronics design engineer is to find ways to compensate for or to minimize these effects, while always delivering smaller, faster, and cheaper devices.

This book, Microelectronics, provides a wide-reaching, regular, and inclusive update on microelectronic circuits and systems.

CHAPTER 1

A Study of Lateralized Cognitive Processes in Upper-Division Electrical Engineering Students': Correlating Written Language Functions with Analytical Reasoning in Microelectronics

Robert Melendy

Department of Electrical Engineering, Mathematics and Applied Science, College of Engineering, George Fox University, Newberg, USA

ABSTRACT

The human brain is asymmetrical in function, with each of its two hemispheres being somewhat responsible for distinct cognitive and motor tasks, to include writing. It stands to reason that engineering students who have established entrance into their upper-division programs will have demonstrated cognitive proficiency in math and logical operations, abstract and analytical reasoning and language usage, to include writing. In this study the question was asked: is there a correlation between an upper-division electrical engineering students' analytical reasoning ability and their descriptive writing ability? Descriptive writing is taken here to mean a students' ability to identify key physical aspects of a mathematical model and to express—in words—a concise and well-balanced description that demonstrates a deep conceptual understanding of the model. This includes more than a description of the variables or the particular application to an engineering problem; it includes a demonstrated recognition of the basic physics that govern the model, certain limitations (idealizations) inherent in the model, and an understanding of how to make practical experimental measurements to verify the governing physics in the model. A student at this level may demonstrate proficiency in their analytical reasoning skills and hence be capable of correctly solving a given problem. However, this does not guarantee that the same student is skilled in associating equations with their physical meaning on a deep conceptual level or in understanding physical limitations of the equation. Consequently, such a student may demonstrate difficulty in mapping their comprehension of the model into written language that demonstrates a sound conceptual understanding of the governing physics. The findings represent a sample of two independent class sections of Electrical and Computer Engineering junior's first course in Microelectronic Devices and Circuits during fall semesters 2012 and 2013 at a private mid-size university in NW Oregon. A total of three exams were administered to each of the 2012/2013 groups. Correlations between exam scores that students achieved on their descriptive writing of microelectronics phenomena and their analytical problem-solving abilities were examined and found to be quite significant.

Keywords

Neurocognition, Abstract Thinking, Analytical Thinking, Microelectronics, Descriptive Writing Abilities

INTRODUCTION

Cognition is a partial foundation upon which psychomotor skills are manifested [1] -[4] . Cognition may either drive or suppress pending actions [3] . In doing so, cognition enables us to produce adaptable and complex movements referred to as psychomotor skills [5].

One such psychomotor skill is writing. This current study however is not focused on the physiological mechanisms of how the brain and cognition produce or control writing; the focus is on the effects of cognition on writing as it relates to high-level problem-solving. Specifically, the study concentrates on identifying a relation between upper-division electrical engineering and computer engineering (EE/CpE) juniors' analytical reasoning abilities and their descriptive writing ability. Descriptive writing is taken here to mean a students' ability to identify key physical aspects of a mathematical model and to express—in words—a concise and well-balanced description that demonstrates a deep conceptual understanding of the devices, circuitry, and the mathematics that describe the operation of these devices. The initial question is: "is there an observable (clinical) relationship between these students' adeptness in writing descriptions of microelectronic device and circuit phenomena and their analytical reasoning (problem solving) skills?" The latter is associated with the cognitive domain [6] ; the former is a psychomotor response [7] (the motor execution of articulation).

CONCEPTUAL FRAMEWORK

Many cognitive models reported in the literature tend to define writing in terms of problem-solving [8] [9]. In this context, it follows that writing arises from an individual's attempt to map language from their thoughts (cognitive domain) onto paper via the formation of written words (a psychomotor process). Summarizing this from a neurocognitive perspective, the former is primarily a phenomenon of the frontal lobes and the latter a phenomenon of the parietal and temporal lobes [4] (Figure 1). Although these neuro-phenomena are controlled by different parts of the brain, education researchers often lump these into cognitive attributes that can be observed as behaviors, for instance, the classification of behaviors that make up Bloom's taxonomy, which is also referred to as Bloom's cognitive domain [10] .

Bloom's cognitive domain informs us that there are six classifications of learning objectives or levels. Using written language to express one's knowledge (Level 1); comprehension (Level 2); application (Level 3); analysis (Level 4); synthesis (Level 5); and evaluation of ideas or information (Level 6). Bloom's taxonomy serves to emphasize one's development of intellectual skills. For example, key words used or the identification of key concepts by "the writer" helps the writer to establish and encourage critical thinking skills [11] [12] . This suggests that a student who can map conceptual ideas from their working memory into writing may also demonstrate conceptual comprehension of the task at hand [13] [14] . In the context of this study, this might measure the ability of an EE/CpE student to map their conceptual ideas of physical semiconductor principles into words that demonstrate comprehension of these principles. What's more, this can range from strategic considerations (such as the organization of physical ideas) to the implementation of motor plans [15] . For instance, choosing and writing key words to describe the physical abstractions within a mathematical model of a semiconductor device.

However, Bloom's cognitive domain does not subcategorize for psychomotor skills, such as describing the

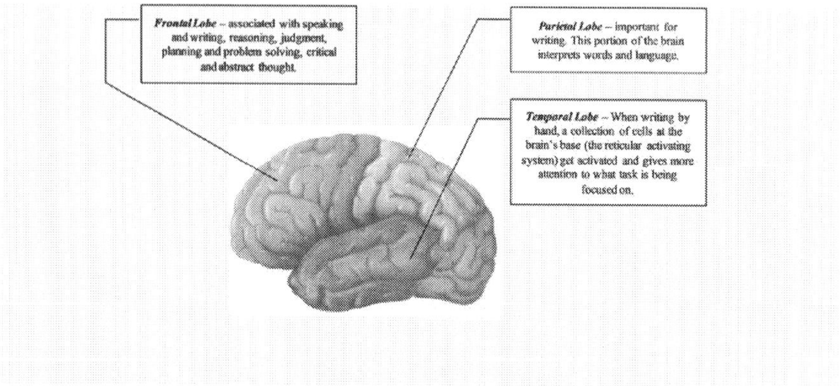

Figure 1.Those two portions of the human brain that are primarily responsible for higher-thought (cognitve domain) and writing (a psychomotor penomenon).

ability of a person to physically manipulate a pencil. These are processes that get described in the field of neurophysiology [16]. A question is therefore: "is there a method to observe one's development of intellectual skills (cognitive domain) via psychomotor responses?" Simpson et al. [17] -[19] developed a "psychomotor taxonomy" in an effort to explain the behavior of a typical learner such as perception (the ability to respond to sensory cues to guide motor activity), responding to phenomena (via explaining and/or showing), and complex overt responses (i.e., where the learner displays mastery in explaining difficult concepts). Bereiter and Scardamalia [20] proposed that individuals who demonstrate skill in mapping thoughts into written language tend to "problematize" a writing task; adopting a strategy they called knowledge transforming. In contrast, a writer who is lower on Bloom's cognitive domain (e.g., Level 1) will typically take a "knowledge-telling" approach, in which written content is generated through association, with one idea prompting the next.

Problem solving has been conceptualized in terms of information processing. Hayes [21] and Flower attempted to classify various activities that occur during writing and the relationships of these activities to: (1) the task environment; (2) the internal knowledge state of the individual. Hayes and Flower suggest that an individual's long-term memory (an aspect of cognitive capacity) has various types of knowledge which not only account for an individual's knowledge of a topic, but a stored plan for how they will demonstrate skill in mapping their thoughts into written, planned content. Hayes and Flower contend that writing involves complex problem solving, in which information is processed by a system of function-specific components in the brain [22] [23].

In a revised model Hayes [24] suggests that different aspects of working memory are inherent in the cognitive process of expressing written language. There is evidence of a correlation between an individual's working memory capacity and their ability to perform analytical reasoning tasks [22] . Given the high levels of cognitive demand that skillful writing involves [25] , individuals who possess well-organized knowledge of a domain and concomitant interest in the subject areas have been shown to demonstrate competence in mapping their comprehension of subject matter into skilled, written language. Furthermore, studies suggest that skilled reasoning in mapping one's comprehension of a topic into written language is associated with analytical reasoning [26] [27].

What's more, planning what knowledge to map into written language and the level of skill in the actual writing process has been found to depend upon one being able to recall relevant knowledge and manipulate it in working memory. For instance, Bereiter [20] and Scardamalia showed a correlation between writing quality and performance on tasks measuring working-memory capacity. Hambrick [28] reported that recall in a readingcomprehension task is facilitated both by high working-memory capacity and high levels of domain knowledge. Similar interactions appear to take place in planning and generating what words to write, where it has been argued that long-term knowledge activated by concepts in working memory functions (in effect) as long-term working memory, thus expanding the set of readily accessible concepts.

If a student is asked to present factual material via writing, some degree of natural conceptual organization is intrinsic to the material to be presented. For instance, to skillfully describe the physical operation of a MOSFET, one could argue that the natural approach would be for the student to begin organizing the written "presentation" in terms of the nand p-type layers and the electron-hole interactions in the presence of an electric field at the gate terminal. This suggests that students who experience difficulty inferring conceptual relationships when learning complex subject matter are likely also to have trouble inferring those relationships when asked to map their conceptualizations into structured, knowledgeable writing.

If this conceptual framework is applied to the study of upper-division EE/CpE students, then in summary, one might predict the following synopsis: Let's say an EE/CpE junior demonstrates adeptness at mapping their comprehension of physical principles and the equations that govern these principles into written language. Let's further say that this student demonstrates objective comprehension and validation of their ideas through written language. It is reasonable to predict that this student should likewise demonstrate effectiveness in executing analytical reasoning problem-solving tasks. Hence, the hypothesis underlying this study is now stated.

HYPOTHESIS

Those EE/CpE juniors who demonstrate adeptness at mapping their comprehension of physical-based equations into written language that demonstrates objective comprehension and validation of their ideas should also demonstrate skill in executing analytical reasoning problem-solving tasks. Put another way, students that restrict their writing to "knowledge-telling" of physical relationships should be less adept at demonstrating objective comprehension of key physical principles or validation of their ideas via problem solving.

DESIGN/METHOD

This study was conducted in two independent sections of Microelectronic Devices and Circuits during the fall semesters of 2012 and 2013 at a private university is NW Oregon. This course consists of three 50 minute weekly lectures and one 150 minute weekly lab. The students in this course are exclusively electrical or computer engineering majors (EE/CpE) at the junior level. The total sample for this study consisted of n = 25 students (22 males and 3 females).

The first step was to assess students' adeptness at mapping their comprehension of physical-based equations into written language. The second step was to assess students' quantitative and analytical problem-solving skills. This was accomplished over the course of three written examinations over the duration of each of the 2012 and 2013 fall semesters. Each exam was divided into two parts: (1) qualitatively describing Microelectronic Equations and Relations (a 20 - 30 minute closed-book and closed-note exam), immediately followed by; (2) quantitative analysis of microelectronic circuits (a 20 - 30 minute exam with a provided equation sheet).

Students were given a 30 minute review two days prior to each exam. They were informed of the exam expectations and that each exam would consist of two parts. They were provided with example problems. So for Part 1, they were shown an example microelectronics describing equation. The expectation was emphasized that they were not to do a literal translation symbol-by-symbol of the equation: that they were expected to demonstrate a conceptual understanding of the equations and the underlying physics. For instance, the following problem was presented on Exam #2: Consider the equation

$$i = I_S\left[e^{\left(\frac{v}{nV_T}\right)} - 1\right]$$

a. (4) What microelectronic circuit model or device does this equation apply to?

b. (10) Describe what each of the variables are in this model and indicate their units:

Variable	Description	Units
I		
I_S		
V		
N		
V_T		

c. (4) In your own words, explain what this equation means.

d. (4) Draw a circuit diagram of the physical situation where this equation would apply.

Label all circuit parameters.

Two-to-three writing problems like this were given on each of the exams. Each student's response was scored as follows:

4 points: student demonstrates deep conceptual meaning of the underlying physics and model limitations.

3 points: student demonstrates some understanding of the underlying physics and/or the model limitations.

2 points: student provides a literal answer, where symbols are identified by correct physical quantities.

1 point: an incomplete or an incorrect answer 0 points: no answer.

Here are sample writings of student responses (Part c.) that were scored using this scale:

4 points: This equation models the current that flows through a non-ideal forward-biased pn-junction diode from an outside source. The current i is dependent on the saturation (or drift) current I_S which is a natural consequence of thermally-generated minority carries that flow from the nto the p-side of the device junction. The current i becomes exponentially large when the voltage v applied to the diode exceeds barrier voltage of the junction. Current I is dependent on the thermal voltage V_T and the ideality factor n, but any variations in these values have little effect on i once the barrier voltage is exceeded.

3 points: This equation tells us that the current flowing in a forward-biased diode is dependent on the saturation current I_S, the thermal voltage V_T, the factor n, and the forward-bias voltage that we apply to the diode. The higher the voltage the higher the current i.

2 points: This equation tells us that the diode current i is equal to saturation current $I_S$multiplied by the exponent of the voltage v divided by n times the thermal voltage minus 1. If i » I_S we can ignore the 1.

1 point: The diode current i depends on many thermal and material property factors such as the saturation current I_S and the thermal voltage V_T and the ideality factor n.

RESULTS AND ANALYSIS

A fundamental question to address is: "is there a meaningful correlation between the students' conceptual writing scores and their performance on the analytical portion of each exam?" Statistically speaking: "is there a spatial pattern among those students who demonstrate adeptness at mapping their comprehension of microelectronic equations into writing and their microelectronic analytical problem solving skills?"

The conceptual written-description score totals from the combined 2012 and 2013 populations were taken as the independent variable; the analytical problem solving score totals were taken as the dependent variable. The first step was to confirm the regression assumption of normality (Figure 2). This plot was done on the residuals and not on the response variables. The justification is that the latter would result in a pseudo-normal probability plot. A pseudo-normal plot would not reflect a correct representation of the population's exam scores. Hence, the initial focus was on the residuals to see if they came from a normal curve. Deviations of the dependent observations (student scores on the analytical portion) from the fitted function are the residuals.

The more linear the distribution of the residuals (i.e., the more a linear function can be fitted to the data) the more faith that the residuals come from a normal curve. As evident in Figure 2 the model's residuals are wellbehaved and are not heteroscedastic. Also, the normal scores are Z-scores. The Z-scores permit the conversion of the independent data sets into scores that can be compared to each other. This analysis method removes arbitrary aspects that may be present in the population and is a widely-accepted statistical tool to check the regression assumption of normality [29] . The conclusion is that the residuals come from a normal distribution.

On the basis of normality a correlation test was done. The rationale: to test for a relationship between the written and analytical exam scores. For the purposes of this study, the goal was merely to show that a relation between these two variables exists, not that the action of achieving a score on the written exam caused a particular score on the analytical portion.

Regression analysis supports that the two exam score variables occur together (Figure 3). Again, the purpose is to provide evidence that the variables are significantly related—not to suggest causality. Arguments made forthe Pearson's correlation coefficient of $r^2 = 0.5854$ being a statistically significant indicator of correlation are discussed as follows.

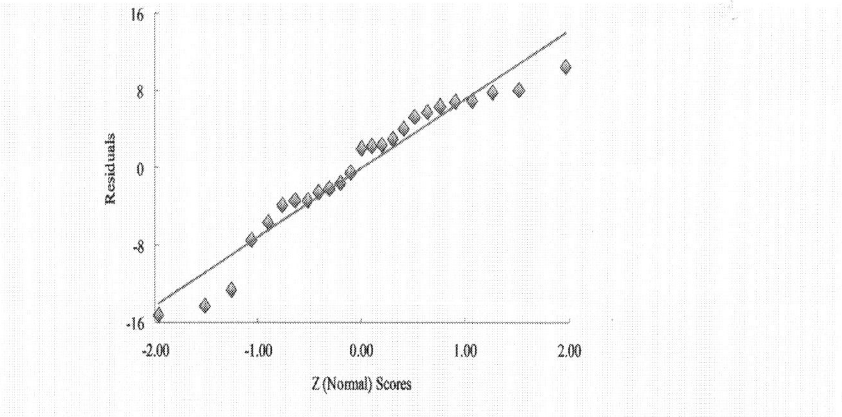

Figure 2. Normal probability plot representing the combined total scores for Microelectronics Exams I, II and III (2012 and 2013 EE/CpE student population). The normality assumption for the sample population tested is supported.

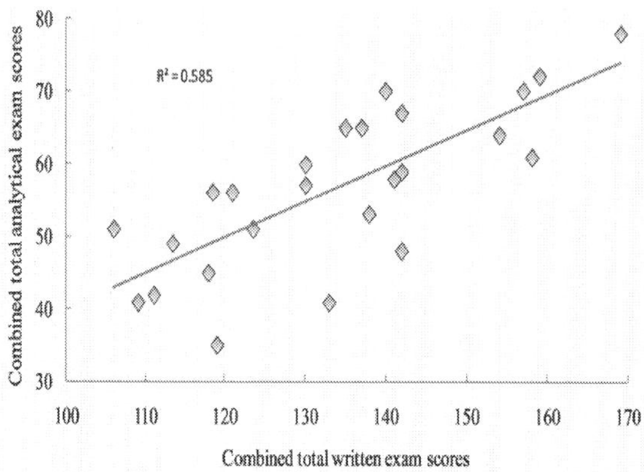

Figure 3. Regression analysis of the combined scores for Microelectronics Exams I, II and III (2012 and 2013 EE/CpE student population).

First, directionality is not an issue. Student's performance on the analytical portion of the exam was measured after they took the written portion of the exam. Thus, the latter could not have possibly caused the former.

Second, it is important to dispense with the existence of a possible third-variable. This would imply the exam scores may be statistically related if there was an unaccounted for variable affecting both sets of exam scores (even in the absence of causality). Thus, to add to the defense of the acceptability of r^2 the question should be asked: "is it possible that a third-variable may have caused the EE/CpE students to score well (or poorly) on both exam portions?" If such a variable exists it may not necessarily be related to one or both exam scores; it is sufficient that the existence of such a variable produces an incidental (spurious) correlation between the scores.

For instance, it can be speculated that the third variable is a students' intelligence quotient (IQ). IQ measures an individual's spatial, mathematical, language and memory abilities [30]. For IQ to be a third variable it would have to cause students to score well (or poorly) on both portions of the exam; either case would produce a statistical relationship between conceptual written explanations and one's analytical problem solving ability. Although it seems reasonable to speculate that IQ may influence one's ability to think analytically and to skillfully communicate their comprehension of difficult concepts, it seems unlikely that IQ would have a consistent effect on one's ability to articulate—via writing—difficult physical and engineering concepts.

Other unaccounted for variables may exists. But just knowing that the written and analytical exam scores are related is interesting, regardless of whether a third-variable exists or not. Rather than continuing to speculate a spurious correlation due to a possible third variable, an additional statistical measure was deemed as a safeguard —the determination of Spearman's rank correlation coefficient, r [31] —a nonparametric measure of statistical dependence between two variables. It assesses how well the relationship between two variables can be described using a monotonic function. When each of the variables is a perfect monotone function of the other, than $r = +1$ or -1. Spearman's rank analysis is warranted since both variables are quantitative and the sample size is relatively small ($n = 25$). Spearman's rank for the population tested is $r = 0.75$ which implies a strong measure of monotonicity between the exam score variables.

CONCLUSIONS AND DISCUSSION

This study focused on determining the existence of a correlation between upper-division EE/CpE students' analytical reasoning abilities and their descriptive writing skills. Descriptive writing was taken to mean a student's ability to identify key physical aspects of a mathematical model and to express—in words—a concise and wellbalanced description that demonstrates a deep conceptual understanding of the model.

Two behavioral components were examined: 1) each student's ability to map their understanding of microelectronics device and circuit models into written language; 2) each student's analytical problem-solving skills of microelectronics device and circuit analysis problems. The initial question asked was: "is there an observable (clinical) relationship between these students' adeptness in writing descriptions of microelectronic device and circuit models and their analytical problem solving skills?" The former is the psychomotor response; the latter is associated with the cognitive domain (refer to Figure 1).

There is convincing evidence that those students who objectively wrote-out knowledgeable and conceptually sound descriptions of semiconductor device and circuit equation models were likewise skillful at solving analytical microelectronics problems. Furthermore, the few students within this population that demonstrated a profound conceptual understanding of microelectronics concepts and provided written validation of their ideas likewise earned the highest-end scores on the analytical portion of their exams.

Regression analysis and a test of monotonicity support a significant correlation between this populations' physical understanding of concepts via writing and solving problems. This seems intuitive—students who perform better on solving analytical-based problems should have more comprehension of the material. The results of the analysis are not meant to imply that the independent variable (writing scores) caused the outcome of the dependent variable (the analytical problem solving scores). It just means there is a statistically significant relationship between the two, thus providing substantial support to the hypothesis (p. 4).

The intellectual merit of this study is that it provides an observable and clinical relationship between EE/CpE junior's adeptness in writing descriptions of microelectronic device and circuit phenomena (students' psychomotor response) and their analytical reasoning (a window into the cognitive domain). Although the former and the latter are lateralized cognitive processes, the research methods carried out in this study enables the engineering educator to research a clinical view into: 1) EE/CpE students' learning behaviors; 2) the level at which EE/CpE juniors are in Bloom's cognitive domain.

The broader impact is this work bridges cognitive science and engineering education in such a way to advance understanding of: 1) the limitations of high-level, analytical problem solving in the absence of students' written evaluation of underlying concepts; 2) Bloom's cognitive domain in the context of a small and unique group of undergraduate students; 3) how to observe engineering students' development of intellectual skills (cognitive domain) through psychomotor responses (writing)—both considered lateralized neurocognitive processes.

SUBSEQUENT RESEARCH

The subsequent research ambitions of the author are to continue to apply this research technique to subsequent groups of upper-division engineering students, not just EE/CpE juniors per sé, but to seniors too and aerospace and mechanical engineering upper-division students as well. With a larger collection of such data across wider population types, perhaps a more thorough understanding into students' lateralized cognitive processes will be realizable. Particularly, the types of cognitive mechanisms that predict how these students store plans for how they'll demonstrate skill in mapping their knowledge of complex subject matter into written language that demonstrates a deep conceptual meaning of the content.

REFERENCES

1. Ackerman, P. and Cianciolo, A. (2000) Cognitive, Perceptual Speed, and Psychomotor Determinants of Individual Differences during Skill Acquisition. Journal of Experimental Psychology: Applied, 4, 259-290. http://dx.doi.org/10.1037/1076-898X.6.4.259
2. Driscoll, M. (2000) Psychology of Learning for Instruction. http://ocw.metu.edu.tr/pluginfile.php/9013/mod_resource/content/1/driscoll-ch10%20(1).pdf
3. Kandel, E., et al. (2000) Principles of Neural Science. McGraw-Hill Health Professions Division, Upper Saddle River, New York, San Francisco, London.
4. Kahol, K., et al. (2008) Effect of Fatigue on Psychomotor and Cognitive Skills. American Journal of Surgery, 195, 195-204. http://dx.doi.org/10.1016/j.amjsurg.2007.10.004
5. Wolpert, D., et al. (2008) The Statistics of Natural Hand Movements. Experimental Brain Research, 188, 223-236. http://dx.doi.org/10.1007/s00221-008-1355-3
6. Driscoll, M. (2001) Psychology of Learning for Assessment. 2nd Edition, Allyn & Bacon, Boston.
7. Amaral, L. and Detmar, M. (2008) From Recording Linguistic Competence to Supporting Inferences about Language Acquisition in Context. Journal of Computer Assisted Language Learning, 21, 323-338. http://dx.doi.org/10.1080/09588220802343454
8. Thevenot, C. and Castel, C. (2012) Relationship and Transfer between Mental and Written Arithmetic. Journal of Cognitive Psychology, 24, 286-294.http://dx.doi.org/10.1080/20445911.2011.617302
9. McCutchen, D. (1986) Domain Knowledge and Linguistic Knowledge in the Development of Writing Ability. Journal of Memory and Language, 25, 431-444.http://dx.doi.org/10.1016/0749-596X(86)90036-7
10. Bloom's Taxonomy. http://classweb.gmu.edu/ndabbagh/Resources/Resources 2/bloomstax.htm
11. Dauer, F. (1989) Critical Thinking: An Introduction to Reasoning. Oxford University Press, New York.
12. Fisher, A. and Scriven, M. (1997) Critical Thinking: Its Definition and Assessment. University of East Anglia, Centre for Research in Critical Thinking.
13. Domjan, M. and Burkhard, B. (1986) The Principles of Learning and Behavior. Brooks/Cole Publishing Co., Belmont.
14. Milner, B., et al. (1988) Cognitive Neuroscience and the Study of Memory. Neuron, 20, 445-468. http://dx.doi.org/10.1016/S0896-6273(00)80987-3

15. Krakauer, J. and Chez, C. (2000). The Organization of Movement. Journal of Computer Assisted Language Learning, 21, 323-338.
16. Knoblich, G., et al. (2002) Authorship Effects in the Prediction of Handwriting Strokes: Evidence for Action Stimulation during Action Perception. The Quarterly Journal of Experimental Psychology Section A: Human Experimental Psychology, 55, 1027-1046.http://dx.doi.org/10.1080/02724980143000631
17. Simpson, E. (1972) The Classification of Educational Objectives in the Psychomotor Domain. Gryphon House, Washington DC.
18. Harrow, A. (1972) A Taxonomy of the Psychomotor Domain. David McKay Publishing Co., New York.
19. Dave, R. (1967) Introduction to the Psychomotor Behaviors. http://en.wikiversity.org/wiki/Introduction_to_the_Psychomotor_Behaviors
20. Bereiter, C. and Scardamalia, M. (1987) The Psychology of Written Composition. Lawrence Erlbaum, Hillsdale.
21. Hayes, J. and Flower, L. (1987) Identifying the Organization of Writing Processes. In: Gregg, L. and Steinberg, E., Eds., Cognitive Processes in Writing, Lawrence Erlbaum, Hillsdale, 3-30.
22. Blakemore, C. (1977) Mechanics of the Mind. Cambridge University Press, Cambridge.
23. McClelland, J. and McNaughton, B. (1995)http://psych.stanford.edu/~jlm/papers/McCMcNaughtonOReilly95.pdf
24. Hayes, J. (1996) A New Framework for Understanding Cognition and Affect in Writing. In: Levy, C. and Ransdell, S., Eds., The Science of Writing: Theories, Methods, Individual Differences and Applications, Lawrence Erlbaum, Mahwah, 1-27.
25. Applebee, A. (1984) Writing and Reasoning. Review of Educational Research, 54, 577-596. http://dx.doi.org/10.3102/00346543054004577
26. Deacon, T. (1997) The Symbolic Species: The Co-Evolution of Language and the Brain. Norton, New York.
27. Dronkers, N. (2000) The Pursuit of Brain-Language Relationships. Brain and Language, 71, 59-61. http://dx.doi.org/10.1006/brln.1999.2212
28. Hambrick, D. (2001) Effects of Domain Knowledge, Working Memory Capacity and Age on Cognitive Performance: An Investigation of the Knowledge-Is-Power Hypothesis. Cognitive Psychology, 44, 339-387. http://dx.doi.org/10.1006/cogp.2001.0769
29. Poole, M. and O'Farrell, P. (1970) The Assumptions of the Linear Regression Model.http://people.uleth.ca/~towni0/PooleOfarrell71.pdf
30. Goldstein, D. and Britt, T. (1994) Visual-Motor Coordination and Intelligence as Predictors of Reading, Mathematics, and Written Language Ability.http://www.amsciepub.com/doi/pdf/10.2466/pms.1994.78.3.819

31. Spearman's Rank Correlation Coefficient.http://en.wikipedia.org/wiki/Spearman's_rank_correlation_coefficient

CITATION

Melendy, R. (2014) A Study of Lateralized Cognitive Processes in Upper-Division Electrical Engineering Students': Correlating Written Language Functions with Analytical Reasoning in Microelectronics. World Journal of Engineering and Technology, 2, 73-81. doi: 10.4236/wjet.2014.22008.

CHAPTER 2

Advances in Microelectronics for Implantable Medical Devices

Andreas Demosthenous

Department of Electronic and Electrical Engineering, University College London, Torrington Place, London WC1E 7JE, UK

ABSTRACT

Implantable medical devices provide therapy to treat numerous health conditions as well as monitoring and diagnosis. Over the years, the development of these devices has seen remarkable progress thanks to tremendous advances in microelectronics, electrode technology, packaging and signal processing techniques. Many of today's implantable devices use wireless technology to supply power and provide communication. There are many challenges when creating an implantable device. Issues such as reliable and fast bidirectional data communication, efficient power delivery to the implantable circuits, low noise and low power for the recording part of the system, and delivery of safe stimulation to avoid tissue and electrode damage are some of the challenges faced by the microelectronics circuit designer. This paper provides a review of advances in microelectronics over the last decade or so for implantable medical devices and systems. The focus is on neural recording and stimulation circuits suitable for fabrication in modern silicon process technologies and biotelemetry methods for power and data transfer, with particular emphasis on methods employing radio frequency inductive coupling. The paper concludes by highlighting some of the issues that will drive future research in the field.

INTRODUCTION

Neuroengineering, the application of engineering techniques to understand, repair, replace, enhance, or otherwise exploit the properties of neural systems, is a topic that is currently generating considerable interest in the research community. The nervous system is a complex network of neurons and glial cells. It comprises the central nervous system (brain and spinal cord) and the peripheral nervous system. Injuries or diseases that affect the nervous system can result in some of the most devastating medical conditions. Conditions, such as stroke, epilepsy, spinal cord injury, and Parkinson's disease, to name but a few, as well as more general symptoms such as pain and depression have been shown to benefit from implantable medical devices. These devices are used to bypass dysfunctional pathways in the nervous system by applying electronics to replace lost function.

The first implantable medical devices were introduced in the late 1950s with the advent of the heart pacemaker [1,2] and subsequently the cochlear implant [3,4]. Both have restored functionality for hundreds of thousands of patients. A pacemaker uses electronics and sensors to continuously monitor the heart's electrical activity and when arrhythmia is detected, electrical stimulus is applied to the heart (via electrodes) to regulate its speed. A cochlear implant uses electronics to detect and encode sound and then stimulate the auditory nerve to enable deaf individuals to hear. Thanks to remarkable advances in microelectronics, electrode technology, packaging, and biomedical signal processing, active implantable medical devices have developed into advanced systems, employing wireless telemetry for transmission of data and sometimes power.

The success of cochlear implants has inspired the development of implantable devices for restoring other basic human sensations. Visual prosthesis translates camera input into electrical stimulation to the visual nervous system to create pixelized vision [5,6], while vestibular prosthesis connects motion sensors to vestibular nerves to restore balance sensation [7,8]. Another example is deep brain stimulation (DBS) that has been shown to provide therapeutic benefits for otherwise treatment-resistant neurological disorders such as Parkinson's disease, tremor, and dystonia [9]. In current clinical DBS systems, high frequency stimulation (>100 Hz) produced by a pulse generator (stimulator) is continuously applied via deep brain electrodes to the targeted tissue area in the brain (Figure1). The characteristics of these pulses (e.g., frequency, pulse duration, and intensity) are programmed into the stimulator (implanted in the chest area) and adjusted via an external programmer. DBS emerged from heart pacemaker technology and hence is also called a brain pacemaker. Current developments are focused on the design of cranial-mounted inductively-powered DBS systems [10] to reduce the length of the leads from the stimulator to the electrodes and novel stimulation techniques using high-density segmented electrodes, which can enable current-steering and electric field shaping capability [11,12]. In addition, recent papers have reported the development of prototype closed-loop (i.e., sense and stimulate) neuroprosthetic devices for applications such as vestibular prosthesis [13] and epilepsy [14,15]. In the case of epilepsy, the device uses implantable multielectrode arrays and amplifiers to record electrical signals from neurons in the brain. The recorded data is then processed to extract important events, which, for example, predict the onset of an epileptic seizure, and electrical stimulation is applied to inhibit the attack.

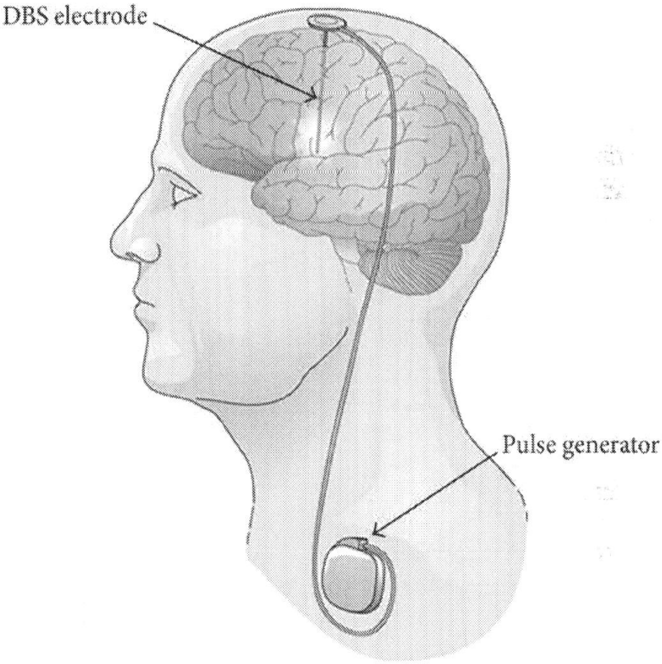

Figure 1. A DBS implant typically consists of an implanted pulse generator which generates electrical pulses for stimulation, a set of connection cables, and an electrode rod which delivers the stimulation pulses to the brain target area. Image source:http://cdn.physorg.com/.

This paper provides a review of advances in microelectronics for implantable medical devices and systems. The focus is limited to neural recording and stimulation circuits suitable for fabrication in modern silicon process technologies and biotelemetry methods for power and data transfer. After this introductory part, Section2reviews several techniques and circuit topologies for neural recording, including methods for on-chip data reduction to reduce the bandwidth requirements of the wireless transmission link. Section3covers advances in neural stimulation circuits and Section4discusses biotelemetry methods including wireless power and data transmission by inductive coupling. Finally, conclusions are drawn in Section5including some suggestions for future research.

NEURAL AMPLIFIERS

Neural signals are low frequency and low amplitude signals. For example, the amplitude of the electroneurogram (ENG) recorded with implanted cuff electrodes [16,17] is typically in the region of 1 μV, with most energy concentrated between 300 Hz and 5 kHz. Even when recording neural activity with penetrating microelectrodes such as the Utah Electrode Array [18,19] or the NeuroNexus penetrating probes (http://www.neuronexustech.com/), the recorded neural action potentials often have amplitudes of only a few tens of microvolts. Hence, circuits for neural signal amplification must have low noise performance and additionally low power consumption so that battery life is prolonged, especially in implantable systems (e.g., implantable loop recorder for long-term monitoring of the heart's electrical activity [20]). In addition, front-end neural amplifiers are required to reject electrode offsets or common-mode interference. Both clock-based and continuous-time techniques have been used in the design of neural amplifiers.

Clock-Based Techniques

The noise in CMOS transistors is usually dominated by flicker (1/f) noise up to relatively high frequencies of the order of several tens of kHz [21]. This is particularly troublesome for the design of low frequency, low noise analog circuits. Typically, p-channel transistors have less 1/f noise than n-channel transistors. Various clock-based techniques have been developed to reduce the effects of1/fnoise. Noise reduction based on physical effects (switched biasing), chopper modulation and autozeroing, is amongst these techniques.

Theswitched biasingtechnique (Figure2) reduces the1/fnoise of a MOS transistor by cyclically increasing and decreasing its gate bias so that the device alternates between strong inversion and accumulation [22]. The transistor noise is modulated by the switching signal. The switching operation is represented as a multiplication of the 1/f noise current (I_{noise}) with the switching signalm(t). For a square-wave signal with 50% duty cycle

$$m(t) = \frac{1}{2} + \frac{2}{\pi}\sin(2\pi f_s t) + \frac{2}{3\pi}\sin(6\pi f_s t) + \frac{2}{5\pi}\sin(10\pi f_s t) + \cdots, \quad (1)$$

where $f_s$is the switching frequency. If $f_s$is set sufficiently high, the baseband noise reduces by half and any modulation effects represented by the sine terms in (1) remain outside the bandwidth of interest and can be removed by filtering. Switched biasing improves noise performance at low frequencies but requires a high speed clock applied to the gate (or bulk) of the transistor with potential problems due to charge feedthrough and additional noise originating from the driver circuit. In addition to reducing the intrinsic 1/f noise, the switched biasing technique reduces power consumption [22]. A switched biasing amplifier demonstrating input-referred noise reduction at low frequencies (<100 Hz) is described in [23]. It uses two operational amplifiers (opamps) configured as buffers linked via resistors. The first stage of each opamp is an operational transconductance amplifier (OTA) circuit of the type shown in Figure2where switched biasing is applied to the p-channel transistor supplying the tail current I_{Bias}. Further noise reduction could be achieved by cycling the voltage bias applied to the bulk terminals of the differential pair input transistors.

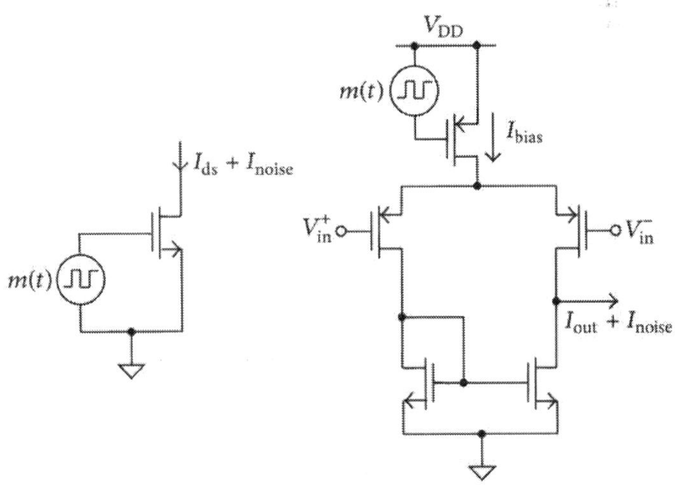

Figure 2. Cycling of MOS gate bias (left) and application to an OTA.

Thechopper techniqueis also based on signal modulation. The technique enables the design of amplifiers with high common-mode rejection ratio (CMRR) performance. A block diagram of the chopper amplifier is shown in Figure3[24]. Before amplification, the amplifier input signal is modulated by a square-wave signal of frequency $f_s$that is much higher than the baseband frequencies of interest. The chopping signal may be represented as in (1). The upconverted signal is then amplified and bandpass filtered. The modulated signal spectrum is located at frequencies higher than the 1/s noise corner. After amplification, the signal is converted back to baseband by multiplication with the same modulation waveform used for upconversion. Lowpass filtering restores the desired signal. The technique reduces both 1/f noise and amplifier dc offset voltages, but the noise performance is ultimately limited by the noise floor of the amplifier. In addition, practical nonidealities, including the finite amplifier bandwidth, can lead to signal distortion.

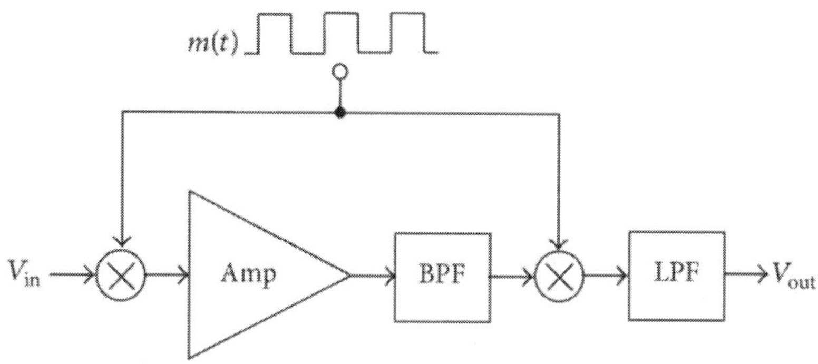

Figure 3. Block diagram of a chopper amplifier.

Several integrated neural amplifiers employing the chopper technique have been described for recording from both implanted and surface electrodes [25–30]. The design in [28] employs an AC-coupled chopping technique to reject electrode offsets and achieve low noise performance. The concept of this technique is shown in Figure4. The system consists of a feedforward stage and a feedback stage. To suppress the 1/f noise of the instrumentation amplifier (IA), the feedforward stage employs an input chopper and an output chopper. To eliminate the electrode offset, the feedback stage employs a lowpass OTA stage followed by a chopper stage. The operation of the circuit is as follows. The input differential electrode offset ((V_{el_offset})) is modulated by the input chopper and appears across resistor R_1. By action of the current-feedback the current through $R_1$is copied to $R_2$and defines the output voltage after demodulation by the output chopper. The lowpass OTA stage filters the dc component of the output and converts it into current. The OTA output current is in turn modulated by a chopper stage. In the steady state, the current supplied by the OTA is V_{el_offset}/R_1. As a result, no current is supplied by the IA and the current passing through $R_2$is zero, so the output (V_{out}) is zero. The neural amplifier in [29] uses chopper modulation and switched-capacitor techniques to reduce the 1/f noise and achieve frequency tuning. The circuit is capable of simultaneous recording of extracellular unit spikes (action potentials) and local field potentials (low frequency signals in the 1 Hz to 100 Hz range). Another publication applies chopper stabilization with a distortion cancelling technique to the design of a front-end transimpedance amplifier for current-mode biosensors [30]. It achieves both low noise and low distortion performance at minimum current consumption. The cancellation technique reduces the distortion and, combined with chopping, substantially reduces the 1/f noise. For a current consumption of 50 μA, it is shown that the input-referred noise density without cancelling and chopping at 10 Hz is 29 pA/√(Hz) and reduces to 3 pA/√(Hz) when both techniques are employed. The total harmonic distortion for a peak current of 1 μA is −55 dB.

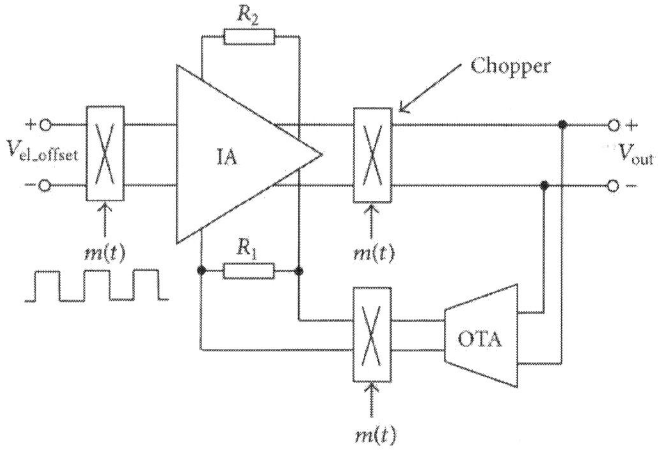

Figure 4. Amplifier with 1/f noise reduction and electrode offset elimination [28].

Theautozeroing techniqueis shown in Figure5. During sampling phase (φ_1), the amplifier is configured as a unity gain buffer and the input noise is sampled. During the amplification phase (φ_2), the noise sample is subtracted from the instantaneous amplifier input noise. As the sampling frequency chosen is higher than the 1/f noise frequency the sample is highly correlated to the instantaneous noise and the low frequency noise is cancelled. A detailed analysis of the autozeroing technique is given in [31]. A drawback of the technique is that high frequency white noise is undersampled and folded back into the baseband where it increases the noise floor. A bandpass micropower neural amplifier employing autozeroing and featuring variable-gain capability is presented in [32]. An interesting design of a low voltage, low noise amplifier combining autozeroing and chopping stabilization is described in [33].

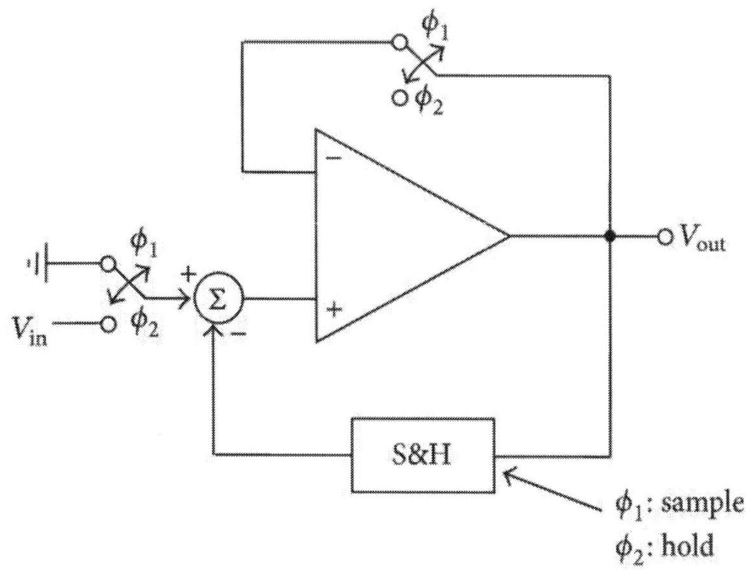

Figure 5. Autozero amplifier.

Continuous-Time Techniques

The noise reduction techniques described above require a clock generation circuit and thus suffer from potential problems associated with high frequency interference and clock feedthrough. In addition, high frequency switching circuits can increase the complexity and power consumption of the design. As an alternative, continuous-time techniques have been extensively used in the design of neural amplifiers. The classic circuit is the ac-coupled OTA-based neural amplifier with capacitive feedback shown in Figure6[34]. The circuit is built around a single stage OTA in CMOS technology. The ratio of capacitors $C_1$and C_2sets the midband gain of the bandpass response. The input is capacitively coupled through, so any dc offset from the electrode-tissue interface is removed (should be made much smaller than the electrode impedance to minimize signal attenuation). Transistors M_a-$M_d$implement MOS pseudoresistors with an extremely large incremental resistance (>10^{12} Ω). This allows the cutoff frequency of the input high-pass filters (i.e., ac-coupled stage) to be set to the millihertz region. The lower cutoff frequency is set by the product of $C_2$and the MOS pseudoresistor implemented by $M_a$and M_b. The upper cutoff frequency is a function of the load capacitance (C_L), the OTA transconductance (G_M), and the midband gain (C_1/C_2). To reduce the effect of 1/f noise, the OTA input transistors should be p-channel devices with large gate areas. Numerous designs of neural amplifier based on the circuit in Figure6or with some variations (e.g., in the realization of the pseudoresistors, use of fully-differential topology with one or two stage OTAs, use of current-reuse techniques to double the transconductance, and so forth [35–39]) have been reported in the literature, including commercial amplifier chips by Intan Technologies, LLC (http://www.intantech.com/). Methods for effective optimization of a recording channel in terms of its power consumption, input-referred voltage noise, silicon area, and technology used are discussed in [40]. The design of nanopower OTAs with enhanced linearity is presented in [41].

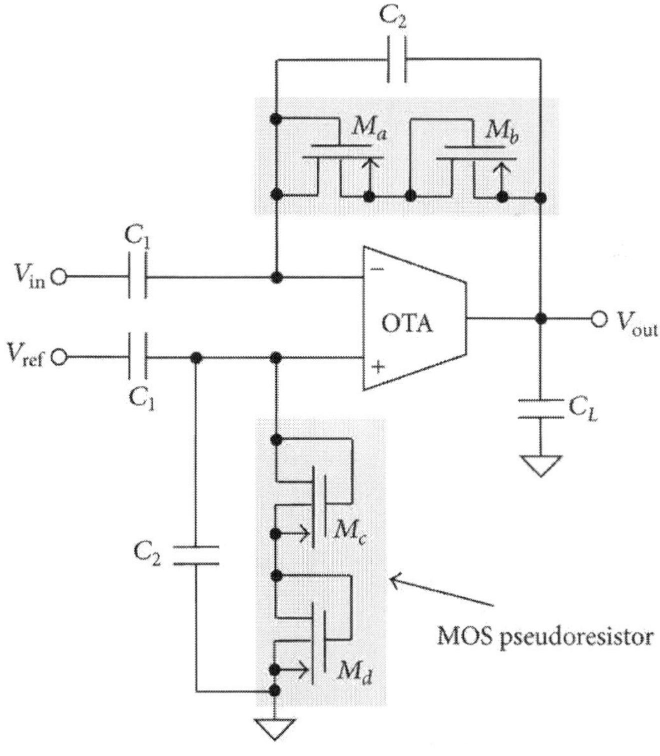

Figure 6. OTA-based amplifier with capacitive feedback [34].

In the case of recording from a multielectrode array, the total power consumption of the amplifier array (as well as the silicon area) may be reduced by using the partial OTA sharing structure proposed in [42]. In this technique, each of the n amplifiers in the array share the components corresponding to the reference electrode (i.e., pseudoresistors $M_c$and $M_d$and capacitors $C_1$and C_2connected to V_{ref}in the amplifier in Figure6). The silicon area is reduced as a benefit of sharing the bulky capacitor C_1. The improvement factor in terms of silicon area depends on the number of shared amplifiers. In general, for an electrode array size of, the OTA sharing technique allows an area saving of $(n-1)/(2n) \times 100\%$, a power saving of $(n-1)/(2n) \times 100\%$, and an improvement in noise efficiency factor (NEF) of $[1-\sqrt{(n+1)/2n}] \times 100\%$, compared to the conventional architecture (i.e., using n neural amplifiers). The NEF is a dimensionless figure of merit (noise-power tradeoff) used to compare different neural amplifier designs. It is defined as [43]

$$\mathrm{NEF} = V_{\mathrm{in,rms}} \sqrt{\frac{2 \cdot I_{\mathrm{tot}}}{\pi \cdot U_T \cdot 4kT \cdot \mathrm{BW}}}, \quad (2)$$

where $V_{\mathrm{in,rms}}$ represents the integrated input-referred noise, $I_{tot}$is the total power consumed by the amplifier, $U_t$is the thermal voltage, k is Boltzmann's constant, t is the absolute temperature in Kelvin, and BW is the −3 dB bandwidth of the amplifier. To include the supply voltage V_{DD}, the modified metric $\mathrm{NEF}^2 \times V_{\mathrm{DD}}$ is used [37].

To achieve low noise performance, a large neural amplifier gain is usually required. The conventional ac-coupled neural amplifier (Figure6) often presents a large input load capacitance (typically 10–20 pF for a midband gain of around 40 dB) to the neural signal source, hence occupying a large silicon area. It suffers from the unavoidable tradeoff between input capacitance and chip area versus the amplifier gain. In the amplifier in Figure7[44], this tradeoff is limited by replacing the feedback capacitor with a clamped T-capacitor network. The diodes are used for discharge purposes. Compared to the conventional circuit this amplifier can achieve a given midband gain with less input capacitance, a higher input impedance and smaller silicon area. For a midband gain of 38.1 dB, a neural amplifier employing the topology in Figure7used a 1.6 pF input capacitance and a total silicon area of 0.056 mm^{2}in 0.35-µm CMOS technology [44].

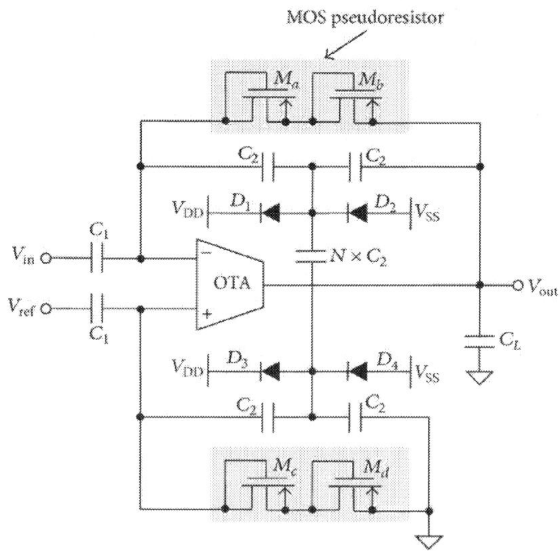

Figure 7. Capacitive feedback amplifier with low input capacitance [44].

Electrode offset removal may also be implemented by means ofactive feedbackloops. An example neural amplifier topology with active feedback for dc rejection is shown in Figure8[45]. It consists of a low noise OTA (OTA_1) with an active feedback circuit implemented by a second OTA (OTA_2) configured as a Miller integrator. The time-constant of the integrator is set by capacitor $C_1$and the MOS pseudoresistor comprising $M_a$and M_b. The midband gain of the amplifier is the same as the gain of OTA_1. The dominant pole of OTA_1sets the amplifier's lowpass cutoff and the common-mode voltage is set by voltage. Another example of a neural amplifier with an active feedback loop to bypass any dc offset current generated by the electrode-tissue interface is described in [46]. It predominantly makes use of current-mode circuit techniques.

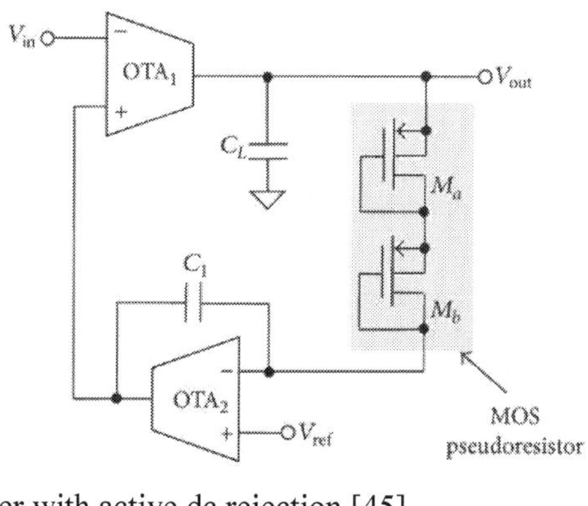

Figure 8. Amplifier with active dc rejection [45].

A low noise neural amplifier for implanted cuff electrodes is described in [47]. The circuit schematic is shown in Figure9. It consists of an input BiCMOS OTA $(Q_1, Q_2, M_1, \text{and } M_2)$ terminated in the load resistor, followed by a first-order bandpass filter (for bandwidth restriction). The upper cutoff frequency is set by the combination of resistor $R_1$and capacitor C_1, while the lower cutoff frequency is set by capacitor $C_1$with the series combination of transistors $M_6$and M_7, the latter transistor pair forming a high value (~20 MΩ) active resistor. In addition to eliminating low frequencies below the pass-band of the input neural signal, the high-pass section of the bandpass filter also removes some of the low frequency flicker noise voltage tail and ensures a dc offset-free amplifier output (V_{out}). The dc bias voltages of $M_6$and $M_7$are provided by the diode-connected transistors $M_8$and M_9, respectively, which are in turn biased by the dc current sources $I_{b2}$and I_{b3}. Circuitry is also included ($M_3$and M_5-Q_3-Q_5) to cancel the base currents ofland. This neural amplifier achieved a measured input-referred root mean square noise voltage of only 290 nV (noise bandwidth of 1 Hz–10 kHz). A variant of this circuit in CMOS technology using lateral bipolar devices forandis possible [48]. In general for a given target noise specification the use of lateral pnp bipolar devices (available as parasitic devices in CMOS technology) tends to require a larger silicon area compared to using standard npn bipolar transistors in BiCMOS technology (but BiCMOS technology might incur higher manufacturing costs). A bipolar transistor structure in CMOS technology featuring high matching characteristics is described in [49].

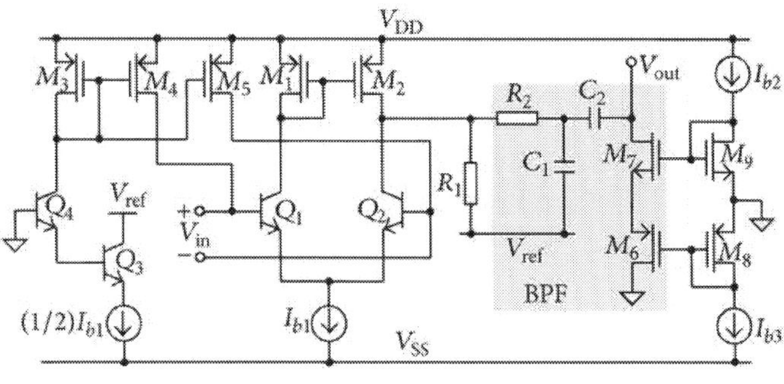

Figure 9. A BiCMOS low noise amplifier [47].

For biomedical front-ends requiring high CMRR performance and accurate gain setting the use of an IA is desirable. An IA may be realized using the classic three-opamp topology. However, the CMRR of the three-opamp IA depends on the matching of the resistors and the need for low output impedance amplifiers can increase power consumption. Another technique for IA design is to employ switched-capacitor circuits [50], but the fold over of noise above the Nyquist frequency can be a major limitation. A popular IA topology for integrated circuits is thccurrcnt feedbacktechnique. In a current feedback IA, the gain is accurately set by the ratio of two resistors and the CMRR does not rely on the matching of resistors. Figure10shows the simplified circuit schematic of a current feedback IA [51]. The input transconductor stage uses a simple current mirror load and current sink biasing. The sensing amplifierserves to exactly balance the drain currents of transistorsandby adjusting the complementary currentsand. A direct result of this is that the input differential voltageis forced across resistorand henceandof the input stage essentially act as a unity-gain buffer. Similarly, the high gain amplifierbalances the drain currents of transistorsandin the output transconductor stage. Since currentsandare exact copies ofand, respectively, the output voltageappears across resistor. Hence, the dc gain of the IA is given by the ratio. Placing capacitorin parallel withcreates a dominant pole, which sets the −3 dB bandwidth of the IA. The CMRR and noise analysis of the circuit are described in [51]. The IA in [52] used the current feedback technique to achieve a CMRR of 99 dB and an input-referred noise of 0.68 μV rms. It was developed to record neural signals from electrodes on the lumbo-sacral nerve roots which can be used to restore lower-body function to patients with paraplegia after spinal cord injury.

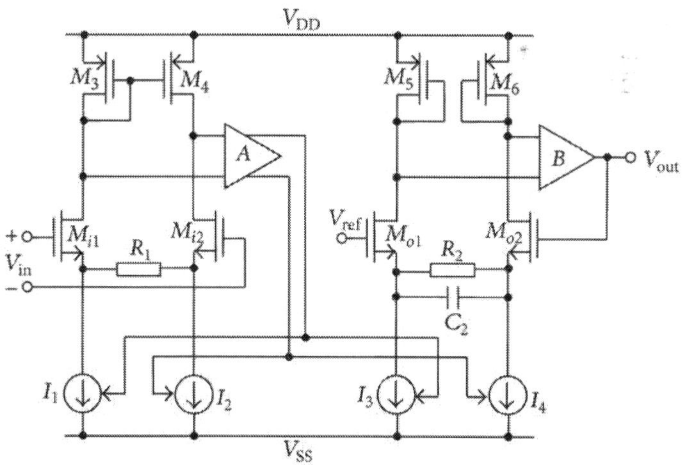

Figure 10. A current feedback instrumentation amplifier [51].

Table1compares various integrated neural amplifiers reported in the literature. From the table, it is observed that there is a relationship between current demand and input-referred noise. In general, the lower the noise performance, the higher the current consumption. There is a wide variation in the silicon area used.

Table 1.Comparison of integrated neural amplifiers.

Parameter	[34]	[109]	[45]	[110]	[111]	[112]	[29]	[42]	[35]	[37]	[44]	[40]	[39]	[46]	[36]
Year	2003	2004	2007	2009	2009	2010	2010	2011	2012	2012	2013	2013	2013	2013	2013
Fully differential	No	No	No	No	Yes	Yes	Yes	No	Yes	Yes	No	No	No	No	Yes
CMOS process (μm)	0.5	1.5	0.18	0.35	0.13	0.35	0.18	0.18	0.13	0.13	0.35	0.18	0.18	0.18	0.18
Area (mm^2)	0.16	0.107	0.05	N/A	N/A	0.02	N/A	0.063	0.054	0.072	0.065	0.065	N/A	0.76	0.16
Supply voltage (V)	5	3	1.8	1	1	3	1.6	1.8	1.2	1	3	1.8	1/1.8	1	1.2
Supply current (μA)	16	38.27	4.67	1.26	12.5	2.8	43.12	4.4	1.6	12.1	2	6.1	1.2	13	0.36
Gain (dB)	39.5	39.3	49.52	45.7	38.3	33	19.1/37.5	39.4	47.5	40	37.5	48/60	54.8/60.9	44.5/55.9	26
Bandwidth (Hz)	25 m–7.2k	0–9.1 k	98.4–9.1k	0.23–7.8k	23 m–11.5k	10–5 k	100–8k	10–7.2k	167–6.9k	50 m–10.5k	1–10k	0.3–9 k	0.38–5.1 k	1–10k	80–15 k
Input-referred noise (μV_{rms})	2.2	7.8	5.6	4.43	1.95	6.08	2.36	3.5	3.8	2.2	10.6	5	4	4.4*	8.1
Noise bandwidth (Hz)	N/A	0.1–10k	1–9.1 k	1–12k	0.1–25.6k	10–5 k	1–8k	10–100k	1–100k	0.1–105k	1–10k	1–8k	1–8k	0.3–10k	N/A
CMRR (dB)	≥83	N/A	52.68	58	>63	60	79	70.1	83	80	74	48	>60	N/A	>60
PSRR (dB)	≥85	N/A	51.93	40	>63	N/A	62	63.8	70	≥80	55	55	>70	N/A	>80
THD Input range (mV_{pp})	1% 16.7	1.1% 5	1% 2.4	0.53% 5.2	1% 1	N/A	N/A	1% 5.7	1% 3.1	1% 1	1% 2.4	1.2 1	1% 1.63	1.03 **	0.05% 10
NEF	4	19.4	4.9	2.16	2.48	5.55	6.68	3.35	2.16	2.9	5.78	4.6	1.9	5.45	1.52
$NEF^2 \times V_{DD}$	80	1129	43.22	4.53	6.15	92.41	71.39	20.20	5.59	8.41	100.22	38.08	3.61	29.7	2.77

*Referred to the input node of the electrode.
**The amplitude range of the input current is $20 nA_{pp}$.

Data Reduction Techniques for Multichannel Neural Recording Front-Ends

Front-end neural recording interfaces for multichannel (multielectrode) systems are typically based on the two types of architecture in Figure11[53]. In the approach in Figure11(a), the analog front-end circuits which amplify and filter the neural signal acquired from each of the electrodes are grouped in an array of channels. Each of these channels comprises a low noise amplifier (LNA) of the type described in Sections2.1and2.2and a bandpass filter, followed by a programmable gain amplifier (PGA) to maximize the output swing. The analog outputs from all the channels are then multiplexed in time and converted into digital words by an analog-to-digital converter (ADC). The generated time-multiplexed data frames can be either digitally processed to compress them or sent directly to the output as raw data. In the approach in Figure11(b), instead of sharing the ADC between the analog outputs of the channels, an ADC is embedded in each channel. Then a common digital processor manages the digitized signals from the channels, classifies (or reduces) the data, and sends it to the output. This solution requires a higher silicon area than the approach in Figure11(a), but it has some merits in terms of power consumption because of the much lower sampling rate requirement at the digitization stage. In addition, the system in Figure11(b)has the advantage that is easily scalable by replicating the channels. A very compact circuit implementation for the topology in Figure11(b)is described in [35], requiring a silicon area of only 0.054 mm^{2}per channel in a 130-nm CMOS process technology.

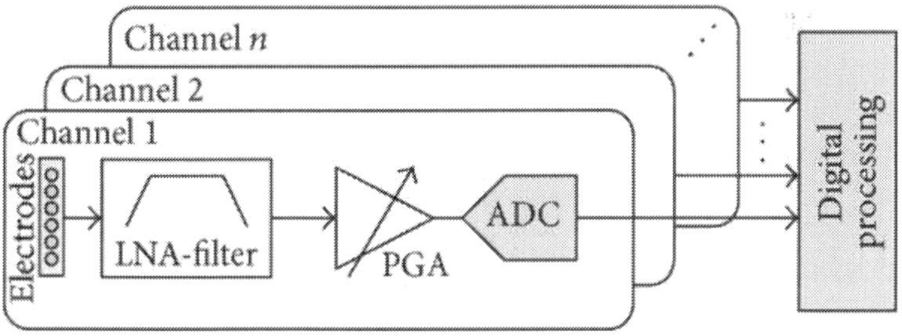

Figure 11. Topologies for multichannel neural recording systems.

Bandwidth limitation is a key issue for wireless biomedical devices. Wireless transmission of raw data is a major challenge for high channel count recording front-ends. For example, a neural interface with 100 channels, a 30 kHz sampling frequency per channel, and an 8-bit sample resolution would generate raw data at 24 Mbps. For a 1024-channel system (for cortical neural sensing), the data rate would increase to a massive 250 Mbps. This data rate is beyond the transmission capabilities of existing (wideband) implantable wireless transmitters [54–56]. In the case of extracellularly recorded action potentials (spikes), spike sorting [57] is an efficient process to achieve on-chip data reduction, thereby enabling lower data rate wireless transmission and low power consumption. Neural recording microsystems with data reduction capability typically employ some sort of thresholding to detect and extract the neural information [58,59]. Within this scheme, detection occurs when the spike's amplitude crosses a specified threshold. The threshold detector can be implemented with analog or digital circuits. For applications where a more detailed classification of multiunit activity into single-unit activity is required, techniques that extract specific biosignal features (e.g., peak-to-peak amplitude, duration, peak-to-zero-crossing time, etc.) are employed. Most spike sorting methods rely on the assumption that each neuron produces a different, distinct shape (as seen by the electrode) that remains constant throughout a recording window. The first step in such techniques isfeature extraction, in which spikes are transformed into a certain set of features that emphasizes the differences between spikes from different neurons as well as the differences between spikes and noise. Thendimensionality reductiontakes place, in which feature coefficients that best separate spikes are identified and stored for subsequent processing, while the rest are discarded. Finally, usingclusteringspikes are classified into different groups, corresponding to different neurons, based on the extracted feature coefficients. Implantable spike sorting hardware must be low power and low area. The algorithms implemented in the hardware must be accurate, automatic, real-time, and computationally efficient. A detailed review and comparison of spike sorting algorithms are provided in [57,60]. An ultra-low power spike sorting digital chip that can perform detection, alignment and feature extraction simultaneously for 64 channels is described in [61]. The chip was implemented in a 90 nm CMOS process and has a power density of 30 $\mu W/mm^2$, which is significantly lower than the power density (800 $\mu W/mm^2$) known to damage brain cells [62].

Spike sorting can potentially reduce the data rate by several orders of magnitude compared to transmitting raw data. However, the data in the segments without spikes (which contains useful information on the neuronal activities) is lost. To preserve these activities, one solution is to use the discrete wavelet transform to process the data before transmission [63]. This allows the retention of almost all of the data but at the cost of chip area and power consumption. An alternative is the emerging field of compressive sensing [64]. Compressive sensing enables signal reconstruction from a small number of nonadaptively acquired sample measurements corresponding to the information content of the signal rather than to its bandwidth. It has simple compression steps and takes advantage of the signal's sparseness, allowing the signal to be determined from relatively few measurements. Energy-efficient compressive sensing methods for implantable neural recording is a topic of current research [65,66].

Table2compares various multichannel neural recording systems with wireless transmission capability. The design in [55] has the highest number of channels (128) and data rate (90 Mbps). The design in [67] has the lowest power consumption per channel.

Table 2.Comparison of wireless multichannel neural recording systems.

Parameter	[113]	[59]	[55]	[114]	[115]	[56]	[67]	[54]	[116]	[15]
Year	2007	2009	2009	2010	2010	2012	2012	2013	2013	2013
Number of channels	100	64	128	64	32	96	64	100	64	64
CMOS process (μm)	0.5	0.5	0.35	0.35	0.5	0.13	0.13	0.5	0.13	0.13
Area (mm^2)	27.73	N/A	63.36	8.37	16.27	25	18.4	25.48	12	12
Supply voltage (V)	3.3	1.8	3.3	3	3	1.2	1.2	3	1.2	1.2/3
Total power (mW)	13.5	14.4	6	17.2	5.85	6.5	0.38	6/90.6	5.03	1.4
Power per channel (μW)	135	225	46.87	269	182.8	67.7	5.93	906	78.5	21.8
Amplifier gain (dB)	40	60	40	65–83	67.8–78	54	47.5–65.5	46	54–60	54–60
Low frequency corner (Hz)	300	10	0.1	1	0.1	280	200	1	10	1
High frequency corner (kHz)	5	9.1	20	10	8	10	6.9	7.8	5	5
Input-referred noise (	5.1	8	4.9	3.05	4.62	2.2	3.8	2.83	6.5	5.1

ADC resolution (bits)	10	8	9	7.2	8.1	10	7.65	12	7.8	7.6
Sampling frequency (kHz)	15	62.5	640	20	N/A	31.25	27/90	20	57	≤100
Downlink modulation	ASK	FSK	N/A	N/A	N/A	N/A	OOK	N/A	OOK	N/A
Carrier frequency (MHz)	2.64	4/8					40.68		915	
Uplink modulation	FSK	OOK	UWB	FSK	FSK	UWB	LSK	FSK	FSK	UWB
Carrier frequency (GHz)	0.433	0.070–0.2	4	0.4	0.915	N/A	0.004	3.2/3.8	0.915	0–1/3.1–10.6
Data rate (Mbps)	0.33	2	90	1.25	0.709	30	1.92	48	1.5	10
On-chip signal processing	Yes	Yes	Yes	Yes	Yes	No	Yes	No	Yes	Yes
Power source	Inductive	Inductive	Battery	Battery	Inductive	Battery	Inductive	Battery	Battery	Battery
Stimulation capability	No	No	No	No	No	No	No	No	No	Yes

NEURAL STIMULATORS

Electrical stimulus applied to nerves can trigger action potentials. At least two electrodes are required in order to produce current flow. The electrodes are commonly arranged in monopolar or bipolar configuration. In both cases, the active (working) electrode is placed near the nerve to be stimulated. In monopolar stimulation, the indifferent electrode is placed away from the active electrode, whilst in bipolar stimulation the reference electrode is placed near the active electrode.

There are two main modes of stimulation, namely, current-mode and voltage-mode, as shown in Figure12, although charge-mode stimulation also exists [68]. Current-mode stimulation (Figure12(a)) is extensively used in implantable stimulators. Active current sources (and sinks) are used to supply the stimulus current to the load (tissue-electrode impedance). For current sources with high output impedance, the stimulus current amplitude is not affected by changes in the load. Examples of integrated current-mode stimulators in CMOS technology for various applications such as nerve root stimulation for spinal cord injury, vestibular prosthesis for balance disorders, and deep brain stimulation for severe movement disorders are described in [11,69–73]. The current amplitude range is from 1 mA to 16 mA.

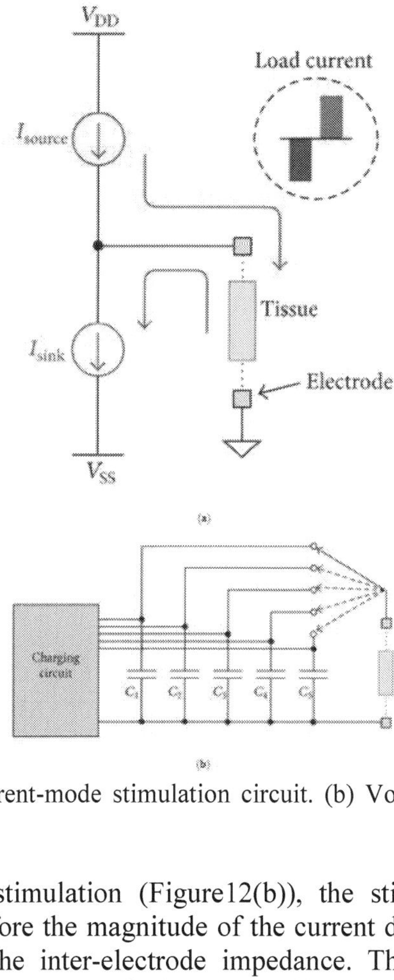

Figure 12. (a) Current-mode stimulation circuit. (b) Voltage-mode stimulation circuit.

In voltage-mode stimulation (Figure12(b)), the stimulator output is a voltage, and therefore the magnitude of the current delivered to the tissue is dependent on the inter-electrode impedance. Thus, it is difficult to control the exact amount of charge supplied to the load because of impedance variations. In the system described in [74], the stimulator drives the electrodes with a sequence of voltage steps, charging the electrode metal-fluid capacitance. This applies a voltage waveform that is an approximation of the waveform that would appear at the electrode if a current pulse was applied (see Figure12(a)). Since the charge is delivered to the electrode directly from the capacitors, it avoids the (substantial) power in the current sources of its current-mode counterpart, as well as providing large voltage compliance. However, this method requires large capacitors (which act as voltage sources) that are difficult to implement on-chip. The design in [74] has five 1 μF capacitors for a 15-electrode stimulation system, which will increase with more electrodes. It is also difficult to achieve fine resolution compared to current-mode stimulation since voltage-mode is an approximate method of producing a current pulse and increasing the resolution requires more capacitors. Another voltage-mode stimulator is described in [75]. Its architecture features energy recovery enabling power savings of 53% to 66% (depending on the load) compared to traditional current-mode stimulator designs.

Common stimulation waveforms are either monophasic or biphasic (Figure13). A monophasic stimulus consists of a repeating unidirectional cathodic pulse (this type of stimulus is common in surface electrode stimulation). A biphasic waveform consists of a repeating current pulse that has a cathodic (negative) phase followed by an anodic (positive) phase. The cathodic phase depolarizes nearby axons and triggers the action potential. The succeeding anodic phase reverses the potentially damaging electrochemical processes that can occur at the electrode-tissue interface during the cathodic phase by (ideally) neutralizing the charge accumulated in the cathodic phase, allowing stimulation without tissue damage. The application of charge-balanced waveforms is very important, especially for implanted electrodes. Usually the stimulus for the cathodic phase is rectangular, supplied by active circuits, while the stimulus for the anodic phase could be either square or exponentially decaying. The rectangular secondary phase is also known as active discharging and the exponentially decaying phase as passive discharging.

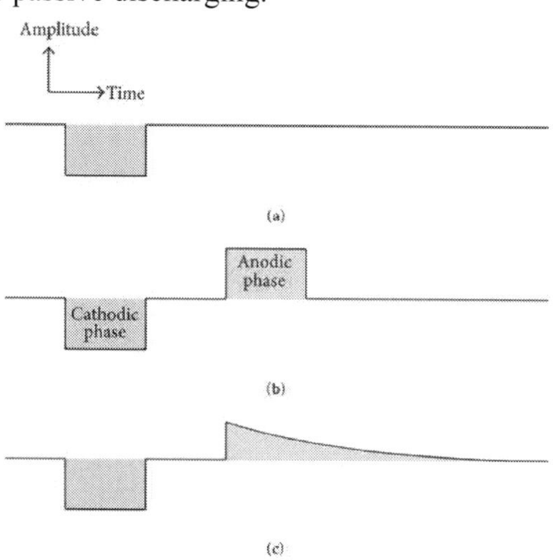

Figure 13. Stimulus waveforms. (a) Monophasic. (b) Biphasic with active cathodic and active anodic phases. (c) Biphasic with active cathodic phase and passive anodic phase (exponential decay).

Charge-Balanced Stimulation

Charge imbalance can be caused by many reasons including semiconductor failure, leakage currents due to crosstalk between adjacent stimulating channels (sites), and cable failure. A blocking capacitor in series with each electrode is used for electrical safety against single-fault conditions [76]. The blocking capacitor is also used to achieve (passive) charge balance. Figure14shows three current-mode stimulator configurations, each employing a blocking capacitor [69]: (a) dual supplies with active phases, (b) single supply with both active phases, and (c) single supply with active cathodic phase and passive anodic phase. The programmable current sinkand current sourcegenerate the cathodic and anodic currents, respectively. These currents are driven through the load by the control of switchesand. When only a single supply is available (Figure14(b)), the anodic and cathodic currents are generated from a single current sink () by reversing the current paths by switch. Both configurations in Figures14(a)and14(b)are (ideally) designed to be charge-balanced to avoid charge accumulation. However, achieving exactly zero net charge after each stimulation cycle is not possible due to mismatch or timing errors and leakage from adjacent stimulating sites. Therefore, it is important to include switchto periodically remove the residual charge by providing an extra passive discharge phase in which the voltage on the blocking capacitor drives current through the electrodes to fully discharge them. Given the necessity for the third phase in the circuits in Figures14(a)and14(b), it is possible to use the passive discharge phase as the main anodic phase as shown in the circuit in Figure14(c)and the corresponding waveform in Figure13(c). Note that the use of capacitive electrodes may also drive the passive discharge phase. However, blocking capacitors may still be deemed necessary to ensure that dc current cannot flow into the electrodes in the event of semiconductor failure due to breakdown voltage or leakage current. Due to the large value required for the blocking capacitors (which can be a few microfarads in the case of stimulators for lower-body applications), they are typically realized as off-chip surface-mount components. For multichannel stimulation implants, a single blocking capacitor per channel may not be sufficient to guarantee safety due to a single-fault failure [77].

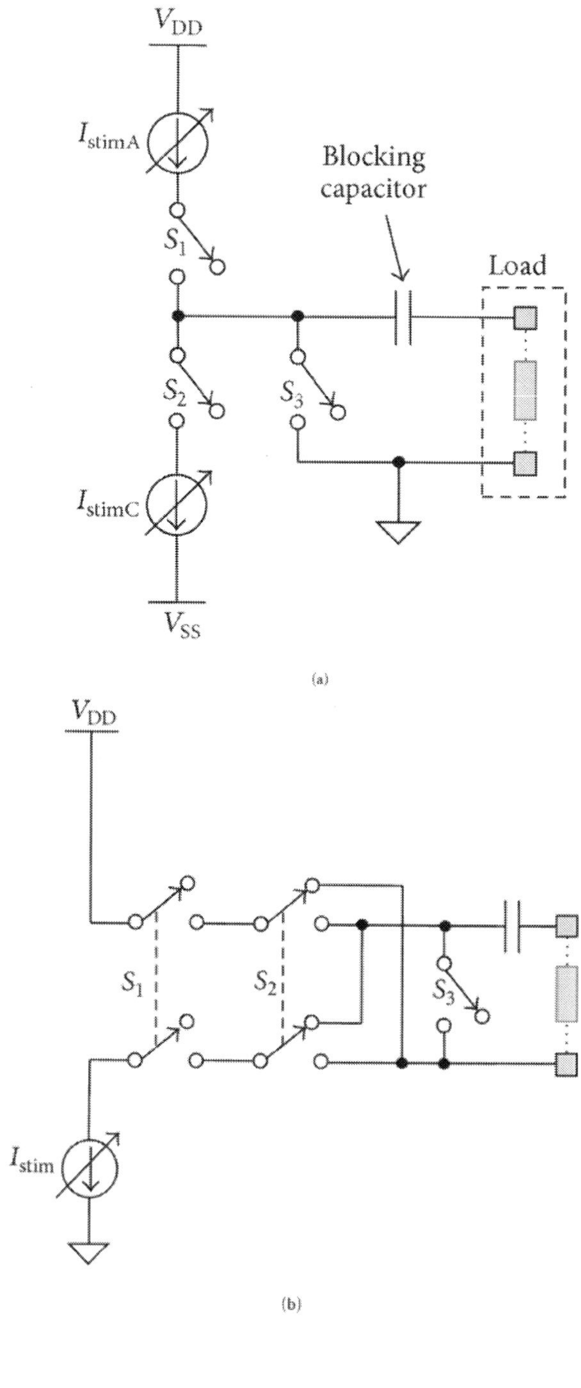

V_{DD}
I_{stimA}
S_1
Blocking capacitor
Load
S_2
S_3
I_{stimC}
V_{SS}
(a)
V_{DD}
S_1
S_2
S_3
I_{stim}
(b)

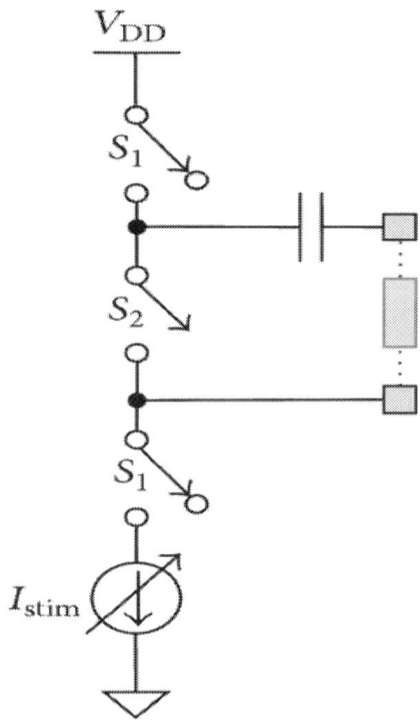

(c)

Figure 14. Current-mode stimulator circuits with blocking capacitor: (a) dual supplies with active cathodic and active anodic phases; (b) Single supply with active cathodic and active anodic phases; (c) single supply with active cathodic phase and passive anodic phase.

For applications where the use of blocking capacitors is not possible due to physical size limitations (e.g., retinal implants with several hundreds of stimulating electrodes), other methods for (active) charge balancing exist (Figure15).(1)Dynamic current balancing [78]. In the circuit in Figure15(a), a current sink is used to generate the cathodic phase and a pMOS transistor () to generate the anodic phase. Before generating a biphasic current pulse,andare open, and the two sampling switches,, are closed. When the circuit settles, the amplitude of the drain current ofis the same as the current sink (due to the feedback), and the resulting bias voltage () on the gate ofis sampled and held. Then the twoswitches open andclose to form the cathodic current. Following thisopens andcloses. Because the gate voltage ofis held at, an anodic current equal to the amplitude of the (cathodic) sink current passes throughto the load. By optimizing the S&H circuit to reduce errors due to charge effects, a residual dc current error of 6 nA is claimed [78].(2)Active charge balancer [79]. In the circuit in Figure15(b), the residual voltage after a biphasic pulse is measured and compared to a safe reference voltage. If the electrode voltage exceeds the safe window, additional short stimulation current pulses are applied to steer the electrode voltage towards a balanced condition. The charge balancer is typically activated for less than 5% of the operation time. This method has the advantage of providing feedback information on the electrode condition after stimulation.(3)H-bridge with multiple current sinks [80]. This approach assumes an asymmetric biphasic waveform. By way of example, consider the H-bridge circuit in Figure15(c). During the high-amplitude cathodic phase, identical current sinksact in parallel to pass current through the electrodes for a time. Then during the low-amplitude anodic phase, one of thecurrent sinks is used to pass current through the electrodes (in the reverse direction) for a time. On a single stimulation cycle this would give inaccurate charge balance as in practice the current sinks would not be perfectly matched. However, if the current sink that is active in the anodic phase is sequentially changed after each stimulus waveform, then aftercycles each sink would have been active for the same amount of time during the cathodic and anodic phases, yielding accurate charge balance. The method is claimed to achieve a maximum charge mismatch of 0.45% [80].(4)Multiphase compensation [71]. The multiphase compensation technique is illustrated by the waveform in Figure15(d). To generate an asymmetric biphasic pulse, the width of the anodic phase is extended totimes the cathodic width, ideally, the anodic current amplitude should betimes smaller than the cathodic amplitude. The amplitude of the currents is controlled through an ADC. Due to the finite resolution of the ADC, the amplitude of the cathodic current, may not be an integer multiple of, whereandis the integer multiple ofthat is the closest to. Thus, there will be charge balance error between the two phases equal. In the multiphase compensation technique, an additional shorter anodic pulse of widthand amplitudeis initiated after the anodic phase to compensate for the error. Subsequently, a passive discharge phase can be applied to further reduce any remaining charge imbalance. The method is claimed to achieve a residual dc current error of 4.5 nA [71].

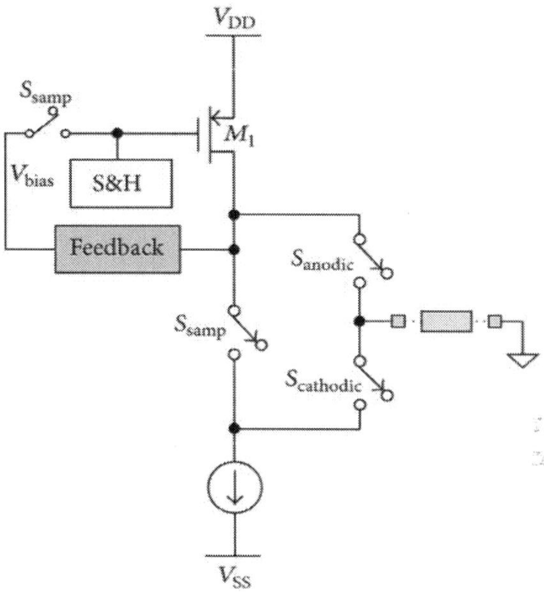
V_DD
S_samp
M_1
V_bias
S&H
Feedback
S_anodic
S_samp
S_cathodic
V_SS

(a)

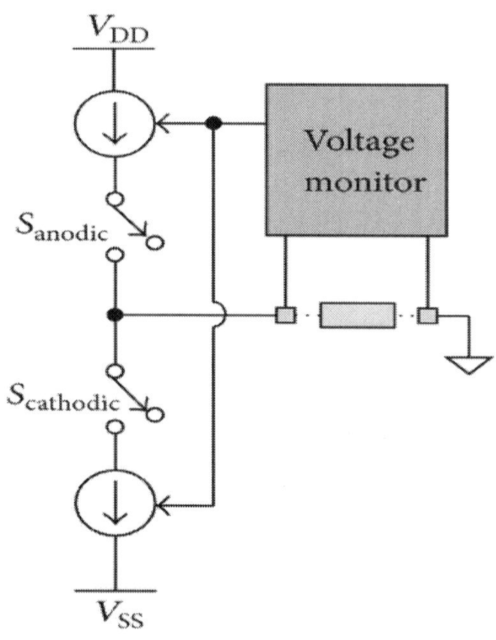
V_DD
Voltage monitor
S_anodic
S_cathodic
V_SS

(b)

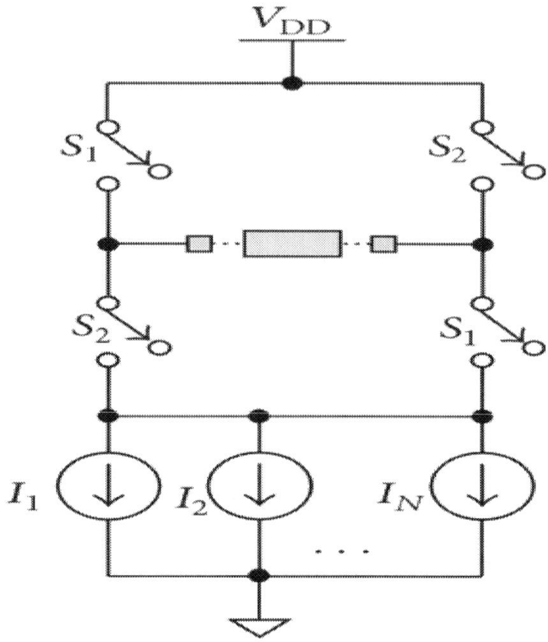

(c)

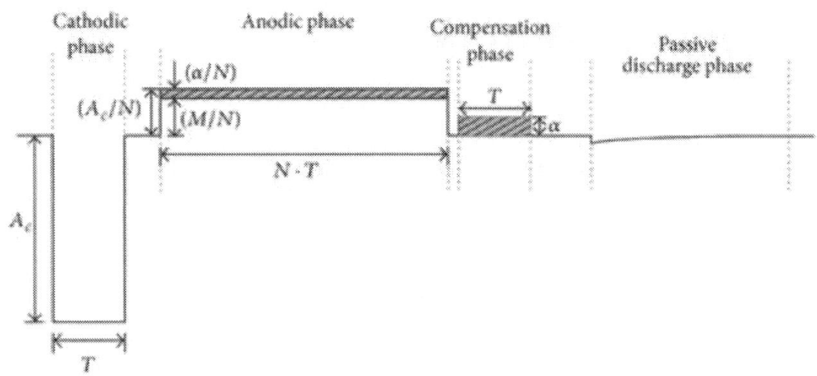

(d)

Figure 15. Techniques for charge balancing. (a) Dynamic current balancing [78]; (b) active charge balancer [79]; (c) H-bridge with multiple current sinks [80]; (d) multiphase compensation approach [71].

Other Stimulator Circuits

A stimulator circuit that is fail-safe with no off-chip blocking capacitors is shown in Figure16[72]. The circuit generates an active stimulation phase by high frequency current switching (HFCS), followed by a passive discharge phasc. During the active stimulation phase, current(generated by a current generator circuit) is switched alternately through the left and right branches of the charge transfer block (Figure16(b)). This high frequency switching mechanism allows the size of the blocking capacitorsandto be significantly reduced. The circuit operation is as follows. During the low-state of the control signal, switchis closed andis open. In this phase, diodeis reverse biased, diodeis forward biased, and currentflows and charges up. In the same phase, on the right branch of the charge transfer block, the control signalis high, and hence switchis open and, andform a closed path which dischargesto one-diode-drop voltage. During the high-state of,is turned low which causesto be charged up andto be discharged. The complementary high frequency currentsandgenerated during the stimulation phase are summed at the anode of the electrode-tissue load. After the stimulation phase, the load is passively discharged via an ac-coupled discharge switch (Figure16(c)) using depletion-mode transistors(connected in parallel for redundancy) which conduct most of the time. The operation of this circuit is as follows. At each negative edge of the control pulse, negative charge is injected into capacitorsand. On the following positive edge diodeis reverse biased so the charge onis retained (apart from the leakage via resistor), and positive charge is injected intobalancing out the injected negative charge on. When a second negative charge arrives it adds to the charge inand the cycle is repeated. After a couple of cycles enough negative charge is built up onto switch off. While the pulses continue, the gate-source voltage ofremains negative, so they are held off. When the pulses cease the negative charge decays via. An ac-coupled switch is also used for the cathode as shown in Figure16(d). Its operation is exactly the same as described above, except now it is the positive edge that provides the charge to. Implementation of the HFCS technique requires silicon-on-insulator (SOI) technology which features fully-isolated active and passive devices. The design in [72] used the X-FAB XT06 process technology (http://www.xfab.com/) which has trench isolation. HFCS is suitable for stimulus current amplitudes up to about 1 mA. The technique can be employed to greatly reduce the size of high-reliability, multichannel stimulator implants sited close to the target nervous tissue (e.g., in the spinal canal or on the brain surface).

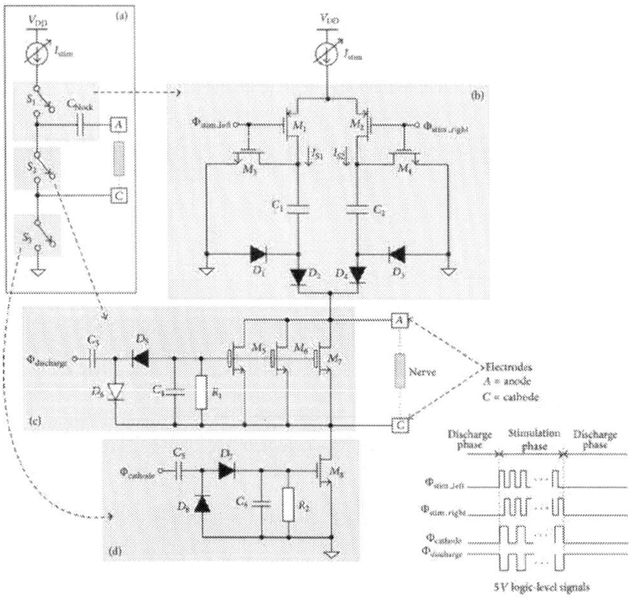

Figure 16. Stimulator circuits [72]: (a) configuration with large off-chip blocking capacitor; (b) HFCS charge transfer block; (c) isolated discharge switch (are depletion transistors); (d) isolated cathode switch. The combination of (b), (c), and (d) provides a safe stimulator circuit with no off-chip blocking capacitors.

There is a demand for high efficiency stimulators in applications which have limited power availability. The basic current-mode stimulator is inherently inefficient and attempts have been made to increase stimulator efficiency. A design is described in [81] which achieves a 2x to 3x reduction in energy consumption compared with the basic current-mode stimulator. It is based on a dc-dc buck voltage converter efficiently providing a variable output for biphasic pulses. The voltage output drive is adjusted by feedback from a sensor detecting the current in the electrode so providing a controlled current independent of the value of the load impedance. The conventional series resistor employed for current sensing wastes energy and is not used. At the output of the dc-dc converter, there is a smoothing capacitor which is in parallel with the electrode load. This capacitor is temporarily disconnected from the converter and the rate of decay of voltage across the capacitor due to the current in the electrode load is detected and used in feedback. This sensing method avoids wasting energy. The system has appreciable sawtooth noise superimposed on the current pulse. Biphasic 400 μA current pulses with widths of 1 ms and rise times of about 100 μs are shown. Components external to the integrated circuit control unit include an inductor (39 μH) and capacitors (up to 10 μF).

Another example of minimizing power dissipation while retaining the basic current-mode generator is presented in [82]. It assumes the use of a high frequency (1 MHz) inductive power link. On its secondary coil, zero voltage switching and adjustment of the conduction angle provide a variable voltage supply to the electrode load and current generator. Feedback adjusts the variable voltage supply so that the current generator operates at just above its compliance limit, minimizing the power it uses. The biphasic current is generated using a single sink current generator with switches similar to Figure15(c). The supply voltage update is near real-time so the quality of the current pulses is high, irrespective of how the load changes during stimulation. Depending on load conditions, 20% to 75% power saving compared to a conventional current-mode stimulator is claimed. In a prototype design, stimulation currents of 20 μA to 1 mA with pulse widths of 20 μs to 200 μs are quoted.

Where cross-coupling between closely spaced, simultaneously operated, stimulating sites must be avoided, floating power supplies are needed. A successful example design, capable of supporting parallel stimulation to electrodes on three semicircular canals for vestibular prosthesis, is described in [83].

DELIVERY OF POWER AND DATA TO IMPLANTS

The requirements imposed on medical devices operating in the body are application specific, but there are a common set of constraints in size, power, and functionality. The interplay between these constraints determines the available processing bandwidth for the electronics, the operating time (in the case of battery operated devices such as pacemakers) and the communication range and bandwidth of the wireless telemetry link. In addition, the location of the device in the body, data rate, frequency, and regulatory standards influence the design complexity and power dissipation of telemetry links.

Long range telemetry links (typically greater than 2 meters) are mainly battery-operated and must conform to strict regulatory standards. They require both high sensitivity receivers and high output power transmitters, both of which result in high power dissipation. In addition, unlike most near-field (short range) inductive links (discussed below) where a stable reference clock can be extracted from the external carrier frequency, long range telemetry links require stable crystal references and frequency synthesizers to generate a local carrier with good frequency stability. These transceivers are typically operated in dedicated frequency bands such as the U.S. federal communications commission (FCC) approved 402–405 MHz for medical implant communication service (MICS) band (e.g., the ZL70102 Microsemi transceiver). This band has a 300 kHz maximum bandwidth and a maximum output power of −16 dBm [84] and the data rate is limited to about 200 kbps. The industrial, scientific, and medical (ISM) radio bands are also frequently used for medical telemetry transmitters. These include the 902–928 MHz, 2.4–2.4835 GHz, and 5.725–5.875 GHz frequency bands and have transmission ranges up to 10 meters. Lower frequencies require larger antennas, while higher frequencies have higher losses due to tissue absorption. The optimum frequency band for wireless transmitters located in the body is reported to be approximately 900 MHz [85]. Ultra wideband (UWB) is an alternative wireless data transmission method used at very low energy levels for short range, high bandwidth communications. Recently, UWB communication in the 3.1–10.6 GHz band was used to develop low power wireless transmitters in CMOS technology for implantable medical devices [15,55,56] for high date rate (10–90 Mbps) transmission from the implant to the external device. Modulation schemes such as on/off keying (OOK) and pulse position modulation (PPM) are used to generate the short pulses. For short range links (typically up to a few centimetres), low frequency inductive links are used for both power supply and bidirectional data transmission. Examples include cochlear and vestibular implants [3,86]. These near-field systems can be made small and highly integrated. Modulation schemes such as OOK, amplitude shift keying (ASK), and frequency shift keying (FSK) are often used for data transmission from the external unit to the implanted device [87,88]. To minimize radio frequency heating due to tissue absorption, these systems are typically operated below 15 MHz with an achievable data rate up to a few Mbps. Transferred power to the implant typically ranges from 10 mW to 125 mW [89–91]. The power transmitted through the tissue should comply with safety standards.

A basic block diagram of a typical architecture for transmitting power/data to an implant via an inductive link is shown in Figure17. The external transmitter typically consists of a class D or class E power amplifier capable of providing large currents in the tuned primary coil () from a relatively low voltage. In the implant, the induced voltage that appears across the secondary coil (, tuned by a capacitor) is rectified and regulated to provide a power supply for the electronics. The data link from the external transmitter to the implant (the downlink) is often achieved by modulating the envelope of the power carrier to create detectable changes across the secondary coil. The data link from the implant to the external circuit (the uplink) is commonly implemented by load modulation techniques. These techniques utilize the property of the coupled coils in which a change in the load of the secondary circuit is reflected back as changing impedance in the primary, through their mutual inductance. Examples of modulation techniques for uplink and downlink data transmission are discussed later.

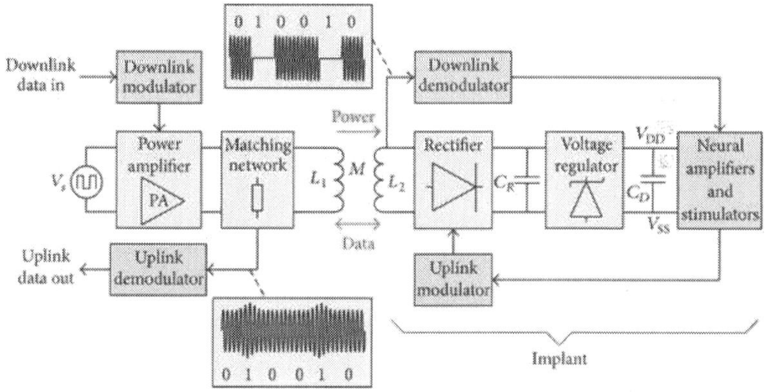

Figure 17. Architecture of a wireless inductive telemetry system to transmit power to an implant and data to and from the implant using a single pair of coils.

The basic equivalent circuit of the inductive link is shown in Figure18[92,93]. The primary circuit („) is tuned in series in order to provide a low impedance load to the driving transmitter. Resistorincludes the loss of the inductorand the output resistance of the voltage source(representing the transmitter driver). At resonance, the voltages related withandcancel each other and thus the primary circuit requires small voltage swings at its inputs. Thus, the primary circuit loads the secondary circuit with a small load. The topology of the secondary circuit is tuned in parallel in order to amplify sufficiently the induced voltage to drive a nonlinear (rectifier) load. Resistorincludes the loss of the inductorand the load resistance of the implant circuits. Both RLC circuits are tuned to the same resonant frequency. The gain factor of the link at resonance is [93]

$$\left|\frac{V_2}{V_s}\right| = \frac{\sqrt{(R_2/R_1)}}{((k_{crit}/k) + (k/k_{crit}))}; \quad k_{crit} = \sqrt{\frac{C_1 R_1}{C_2 R_2}}, \tag{3}$$

whereis the coupling coefficient andis the critical coupling coefficient. The relative dimensions of the coils and the air gap cause a low coupling coefficient (). In addition, the inductances of the primaryand secondarycoils are small and can be generated using coils with a few turns. The typical variations of the gain factor are illustrated in Figure19. The gain factor is not constant with respect to coupling variations. The link transfers maximum power when the resistance in the primary is equivalent to the reflected secondary resistance, assuming that reactive components are being cancelled. It can be shown that efficiency () at critical coupling is equal to 50% (Figure20) [93]. For the purposes of data transfer, the link behaves as a narrow-band bandpass filter. At resonance and assuming that the coils have the same quality factor, the bandwidth of the link iswhereis the carrier frequency [87].

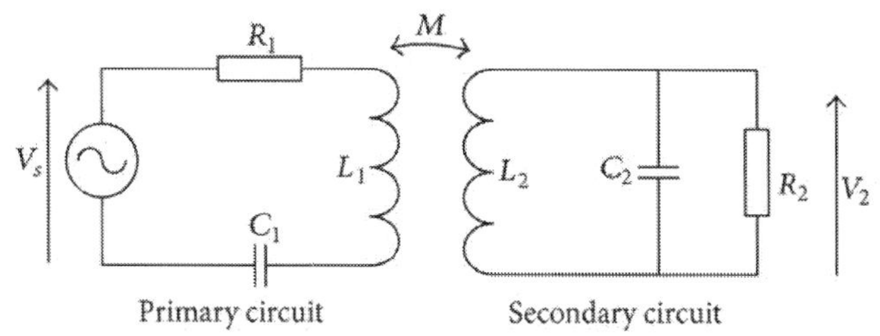

Figure 18. Equivalent circuit of the inductive link with a load ().

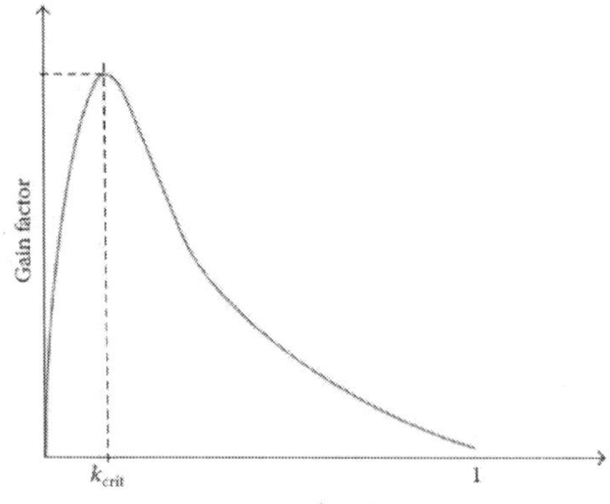

Figure 19. Gain factor versus coupling when the primary and secondary circuits are tuned to the same frequency.

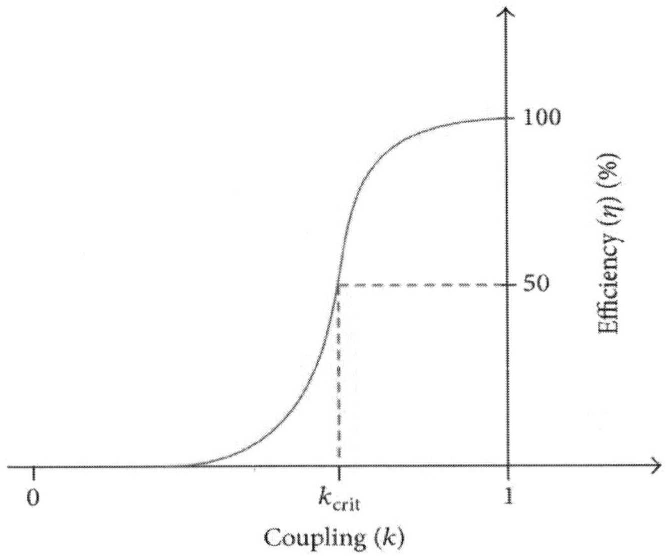

Figure 20. Efficiency of the inductive link versus coupling when the primary and secondary circuits are tuned to the same frequency.

Power transfer (gain factor) optimization and data transfer (bit rate) optimization have contradicting requirements. To illustrate this, the circuit in Figure18was simulated in Advanced Design System (Agilent EEsof EDA) with (Ω, μH) and (kΩ, μH) tuned to 1 MHz byand, respectively [94]. The relationships between gain factor (power transfer), carrier frequency, and coupling coefficient are plotted in Figure21. The optimum gain factor is at the resonant frequency (1 MHz) and coupling coefficient () of around 0.05. The gain factor can be improved by increasing the quality factors. The bandwidth of the inductive link can be increased by lowering the quality factors. It also increases as the coupling coefficient increases (i.e., the gap between the coils decreases). In practice, the inductive link should be designed to account for the potentially wide variation in the gap between the coils. Coupling compensation techniques (incorporating feedback) regulate the voltage across the secondary coil [95–97]. Power and data links can use separate sets of coils because this allows independent optimization to maximize performance [98]. However, using separate coils has the drawbacks of an increase in the implant footprint and electromagnetic interference. The latter requires the use of complex modulation techniques to minimize its effects and increases system complexity.

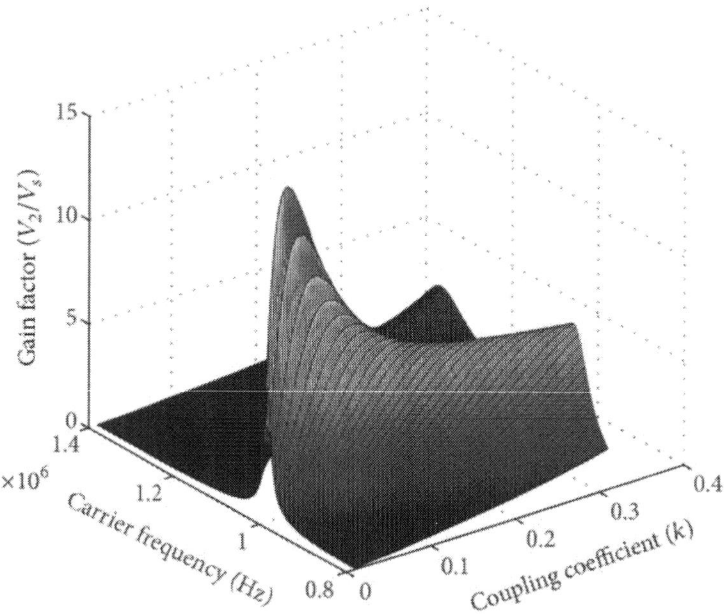

Figure 21. Inductive link gain factor versus coupling coefficient and carrier frequency simulated in ADS for the equivalent circuit model of Figure18.

The most commonly used technique for uplink data transmission is passive signalling [99] also known as load shift keying (LSK) [100]. This modulation is based on the reflection of the implant's load to the transmitter via the inductive link. A typical implementation is shown in Figure22. The binary Data stream shorts the implant coil and the change in impedance is reflected in the transmitter because the implant load is much larger than the on-resistance of the switch transistor. Where only two coils are used (for both power and data) there is a risk of disruption in power delivery if the short is applied for too long to the implant coil. When communication is in idle mode, the link should be optimized for maximum power transfer. The bandwidth of LSK is limited by the coupling factor, the parameters of the coils, and the transient response of the inductive link. Multilevel LSK may be used to increase the data rate [101]. An alternative communication technique for uplink is passive phase shift keying (PPSK) [102]. Unlike LSK, the switch across the secondary coil (Figure22) closes synchronously with the carrier for half the carrier cycle. The transient response in the primary coil current is detected as a logic "1" signal. An integrated implementation of PPSK in CMOS technology is presented in [103]. The circuit was designed to work at 13.56 MHz with a single set of coils for both data transmission and power delivery. The link can reach a data rate of up to 1/16 of the carrier frequency, that is, 847.5 kbps in this case, making it the fastest data rate achieved by a single wireless (inductively coupled) link used simultaneously for power delivery and communication for implantable devices. Another method for uplink transmission is pulse harmonic modulation (PHM). It achieves a data rate of 20 Mbps for 1 cm coil separation using a carrier frequency of 66.7 MHz [104]. A pattern of very narrow pulses with specific time delays and amplitude is used, which minimizes intersymbol interference across the receiver coil. PHM is one of the fastest data transmission methods currently known via inductively coupled coils. Unfortunately, to implement it, a separate power link is required as this method is carrier-less.

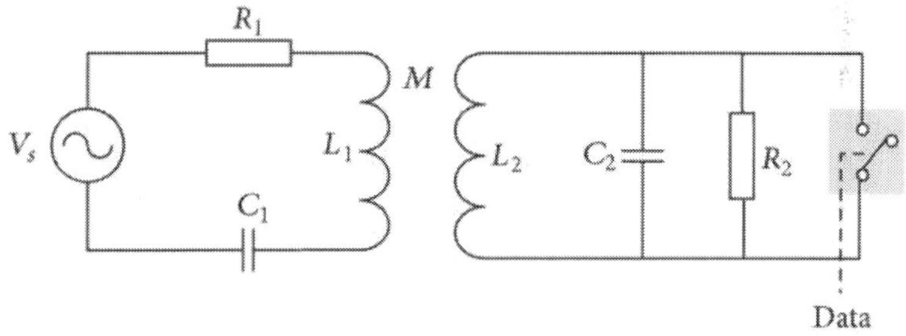

Figure 22. Implementation of LSK modulation.

For downlink data transmission, various digital modulation schemes are used. The most common are binary amplitude shift keying (BASK), binary frequency shift keying (BFSK) and binary phase shift keying (BPSK). The simplest in implementation is BASK but it is sensitive to amplitude fluctuation and the bit rate is typically limited to about 10% of the carrier frequency [87,88]. In BFSK the binary data is represented by constant amplitude, using two different frequencies, where logic "1" is assigned to one frequency and logic "0" to the other. Compared to BASK, BFSK can provide higher data-rate-to-carrier-frequency ratio [105] but it requires a wide passband in the inductive link to allow for the different frequencies, which limits power transfer. The advantage of BPSK over BASK and BFSK is the use of a carrier with fixed amplitude and fixed frequencies [106], enabling efficient and stable power transfer. However, BPSK requires a complicated demodulator, implemented by some kind of phase-locked loop, usually a Costas loop [87]. A comprehensive survey and comparison of modulation techniques is presented in [88].

CONCLUSION AND FUTURE DIRECTIONS

Since the 1950's remarkable efforts have been undertaken in the development of implantable medical devices. Initially most of the successful applications focused on cardiac rhythm management. Today's implantable medical devices provide therapy to treat numerous health conditions. Exciting new applications, for example, in electrical neuromodulation, can be used to treat Parkinson's disease, epilepsy, bladder control, gastrointestinal disorders and numerous psychological disorders such as obsessive-compulsive disorder. Implantable medical devices can now provide a range of pharmacological therapies enabling precise dosage and interval delivery of drugs to more effectively treat patients' conditions while minimizing side effects.

There are many challenges when creating an implantable medical device. These include microelectronic design, electrode technology, packaging, and biomedical signal processing. This paper has focused on the advances in microelectronics over the last decade or so for implantable medical devices and systems. Several examples of neural amplifiers featuring low noise and low power to monitor the small electrical potentials produced from living neurons via electrodes have been discussed. Both clock-based and continuous-time techniques are used in the design of neural amplifiers. Nowadays, implantable neural recording devices include sophisticated signal processing functions on-chip for data compression, such as spike sorting. Analog and mixed analog/digital circuits are used in their implementation. New advanced techniques to further reduce the power and bandwidth requirements of the wireless data transmission link, for example, based on the concept of compressive sensing, will continue to emerge.

Neural stimulation is a key function performed by implantable medical devices for many applications. Some common principles and design techniques for neural stimulators have been presented, including new techniques which avoid the need for off-chip blocking capacitors and methods for achieving charge balancing. There is a need for energy-efficient stimulator circuits to reduce power consumption (two such examples have been discussed) and further developments in this area are anticipated, particularly for applications requiring many stimulation sites (e.g., retinal prosthesis for the blind).

Wireless power and data operation of implantable medical devices are important because they avoid the need for implantable batteries and offer more flexibility to patients. Although major advances have been achieved in the field of wireless communications and wireless powering for implants, further improvements in terms of new techniques that allow better optimization of the entire system are anticipated. The basic principles of inductively coupled telemetry links commonly used to wirelessly power implants and to provide them with a medium for bidirectional communication have been discussed. Recent developments include the introduction of advanced modulation schemes (e.g., PHM) that allow wideband transmission of data over inductively coupled coils. In addition, transceivers based on conventional wideband wireless radio technology are emerging. These are expected to continue to offer improved performance in terms of an increase in output data rate with lower power consumption requirements, as smaller geometry silicon process technologies are used for the implementation of the implantable circuits. On the subject of providing power to implantable medical devices, it is expected that systems using energy harvested from outside and inside the human body will evolve. An example of such a system that uses the cochlear in the inner ear as a battery source to supply a 2.4 GHz radio transmitter is described in [107]. For such systems ultra-low power circuits are essential.

An important research topic involves closed-loop sense and stimulate systems which will continue to develop, possibly combining both sensing of electrical and chemical responses, for applications such as DBS, epilepsy, and other neurological conditions. Such systems will allow better management of the clinical condition allowing systems to adapt to the varying pathological characteristics. In addition, when new highly miniaturized implantable neural interfaces are developed, a step change in micropackaging techniques will be needed to protect the active area of the integrated circuit from hostile environments experienced in the body. As an example, a recent publication describes a technique for integrated circuit micropackages, dedicated to neural interfaces, based on gold-silicon wafer bonding [108].

The expectation of longer life and a progressively increasing knowledge base will place further dependence on modern healthcare technologies to improve the quality of life of a very large number of patients (both young and old). This will undoubtedly provide a fertile ground for future research.

Conflict of Interests

The author declares that there is no conflict of interests regarding the publication of this paper.

REFERENCES

1. C. Ward, S. Henderson, and N. H. Metcalfe, "A short history on pacemakers,"International Journal of Cardiology, vol. 169, no. 4, pp. 244–248, 2013.
2. P. E. Vardas, E. N. Simantirakis, and E. M. Kanoupakis, "New developments in cardiac pacemakers,"Circulation, vol. 127, pp. 2343–2350, 2013.
3. F. G. Zeng, S. Rebscher, W. V. Harrison, X. Sun, and H. Feng, "Cochlear implants: system design, integration, and evaluation,"IEEE Reviews in Biomedical Engineering, vol. 1, pp. 115–142, 2008.
4. B. S. Wilson and M. F. Dorman, "Cochlear implants: a remarkable past and a brilliant future,"Hearing Research, vol. 242, no. 1-2, pp. 3–21, 2008.
5. J. M. Ong and L. da Cruz, "The bionic eye: a review,"Clinical & Experimental Ophthalmology, vol. 40, pp. 6–17, 2012.
6. T. Guenther, N. H. Lovell, and G. J. Suaning, "Bionic vision: system architectures—a review,"Expert Review of Medical Devices, vol. 9, no. 1, pp. 33–48, 2012.
7. C. Wall III, D. M. Merfeld, S. D. Rauch, and F. O. Black, "Vestibular prostheses: the engineering and biomedical issues,"Journal of Vestibular Research: Equilibrium and Orientation, vol. 12, no. 2-3, pp. 95–113, 2002-2003.
8. G. Y. Fridman and C. C. Della Santina, "Progress toward development of a multichannel vestibular prosthesis for treatment of bilateral vestibular deficiency,"Anatomical Record, vol. 295, pp. 2010–2029, 2012.
9. S. Miocinovic, S. Somayajula, S. Chitnis, and J. L. Vitek, "History, applications, and mechanisms of deep brain stimulation,"JAMA Neurology, vol. 70, no. 2, pp. 163–171, 2013.
10. H. M. Lee, H. Park, and M. Ghovanloo, "A power-efficient wireless system with adaptive supply control for deep brain stimulation,"IEEE Journal of Solid-State Circuits, vol. 48, no. 9, pp. 2203–2216, 2013.
11. V. Valente, A. Demosthenous, and R. Bayford, "A tripolar current-steering stimulator ASIC for field shaping in deep brain stimulation,"IEEE Transactions on Biomedical Circuits and Systems, vol. 6, no. 3, pp. 197–207, 2012.
12. V. Valente, A. Demosthenous, and R. Bayford, "Output stage of a current-steering multipolar and multisite deep brain stimulator," inProceedings of the IEEE Biomedical Circuits and Systems Conference (BiOCAS '13), pp. 85–88, Rotterdam, The Netherlands, October-November 2013.
13. J. DiGiovanna, W. Gong, C. Haburcakova et al., "Development of a closed-loop neural prosthesis for vestibular disorders,"Journal of Automatic Control, vol. 20, pp. 27–32, 2010.

14. Berényi, M. Belluscio, D. Mao, and G. Buzsáki, "Closed-loop control of epilepsy by transcranial electrical stimulation,"Science, vol. 337, pp. 735–737, 2012.
15. K. Abdelhalim, H. M. Jafari, L. Kokarovtseva, J. L. Perez Velazquez, and R. Genov, "64-channel UWB wireless neural vector analyzer SOC with a closed-loop phase synchrony-triggered neurostimulator,"IEEE Journal of Solid-State Circuits, vol. 48, no. 10, pp. 2494–2515, 2013.
16. G. E. Loeb and R. A. Peck, "Cuff electrodes for chronic stimulation and recording of peripheral nerve activity,"Journal of Neuroscience Methods, vol. 64, no. 1, pp. 95–103, 1996.
17. R. B. Stein, D. Charles, L. David, J. Jhamandas, A. Mannard, and T. R. Nichols, "Principles underlying new methods for chronic neural recording,"Canadian Journal of Neurological Sciences, vol. 2, no. 3, pp. 235–244, 1975.
18. C. T. Nordhausen, E. M. Maynard, and R. A. Normann, "Single unit recording capabilities of a 100 microelectrode array,"Brain Research, vol. 726, no. 1-2, pp. 129–140, 1996.
19. S. Dickey, A. Suminski, Y. Amit, and N. G. Hatsopoulos, "Single-unit stability using chronically implanted multielectrode arrays,"Journal of Neurophysiology, vol. 102, no. 2, pp. 1331–1339, 2009.
20. S. Mittal, E. Pokushalov, A. Romanov et al., "Long-term ECG monitoring using an implantable loop recorder for the detection of atrial fibrillation after cavotricuspid isthmus ablation in patients with atrial flutter,"Heart Rhythm, vol. 10, no. 11, pp. 1598–1604, 2013.
21. Y. Nemirovsky, I. Brouk, and C. G. Jakobson, "1/fnoise in CMOS transistors for analog applications,"IEEE Transactions on Electron Devices, vol. 48, no. 5, pp. 921–927, 2001.
22. E. A. M. Klumperink, S. L. J. Gierkink, A. P. van der Wel, and B. Nauta, "Reducing MOSFET 1/f noise and power consumption by switched biasing,"IEEE Journal of Solid-State Circuits, vol. 35, no. 7, pp. 994–1001, 2000.
23. Y.-J. Min, C.-K. Kwon, H.-K. Kim, C. Kim, and S.-W. Kim, "A CMOS magnetic hall sensor using a switched biasing amplifier,"IEEE Sensors Journal, vol. 12, no. 5, pp. 1195–1196, 2012.
24. C. C. Enz, E. A. Vittoz, and F. Krummenacher, "A CMOS chopper amplifier,"IEEE Journal of Solid-State Circuits, vol. 22, no. 3, pp. 335–342, 1986.
25. Uranga, X. Navarro, and N. Barniol, "Integrated CMOS amplifier for ENG signal recording,"IEEE Transactions on Biomedical Engineering, vol. 51, no. 12, pp. 2188–2194, 2004.
26. T. Denison, K. Consoer, W. Santa, A.-T. Avestruz, J. Cooley, and A. Kelly, "A2 μw 100 nV/rtHz chopper-stabilized instrumentation amplifier for chronic

measurement of neural field potentials,"IEEE Journal of Solid-State Circuits, vol. 42, no. 12, pp. 2934–2945, 2007.

27. Y. Tseng, Y. Ho, S. Kao, and C. N. I. S. Su, "A 0.09 μ W low power front-end biopotential amplifier for biosignal recording,"IEEE Transactions on Biomedical Circuits and Systems, vol. 6, no. 5, pp. 508–516, 2012.
28. R. F. Yazicioglu, P. Merken, R. Puers, and C. van Hoof, "A 60 μW 60 nV/√Hz readout front-end for portable biopotential acquisition systems,"IEEE Journal of Solid-State Circuits, vol. 42, no. 5, pp. 1100–1110, 2007.
29. J. Lee, M. Johnson, and D. Kipke, "A tunable biquad switched-capacitor amplifier-filter for neural recording,"IEEE Transactions on Biomedical Circuits and Systems, vol. 4, no. 5, pp. 295–300, 2010.
30. V. Balasubramanian, P. F. Ruedi, Y. Temiz, A. Ferretti, C. Guiducci, and C. C. Enz, "A 0.18 μm biosensor front-end based on 1/f noise, distortion cancelation and chopper stabilization techniques,"IEEE Transaction on Biomedical Circuits and Systems, vol. 7, no. 5, pp. 660–673, 2013.
31. C. C. Enz and G. C. Temes, "Circuit techniques for reducing the effects of Op-Amp imperfections: autozeroing, correlated double sampling, and chopper stabilization,"Proceedings of the IEEE, vol. 84, no. 11, pp. 1584–1614, 1996.
32. C.-H. Chan, J. Wills, J. LaCoss, J. J. Granacki, and J. Choma Jr., "A novel variable-gain micro-power band-pass auto-zeroing CMOS amplifier," inProceedings of the IEEE International Symposium on Circuits and Systems (ISCAS '07), pp. 337–340, May 2007.
33. Y. Masui, T. Yoshida, M. Sasaki, and A. Iwata, "0.6 V supply complementary metal oxide semiconductor amplifier using noise reduction technique of autozeroing and chopper stabilization,"Japanese Journal of Applied Physics, vol. 46, no. 4B, pp. 2252–2256, 2007.
34. R. R. Harrison and C. Charles, "A low-power low-noise CMOS amplifier for neural recording applications,"IEEE Journal of Solid-State Circuits, vol. 38, no. 6, pp. 958–965, 2003.
35. Rodríguez-Pérez, J. Ruiz-Amaya, M. Delgado-Restituto, and A. Rodríguez-Vázquez, "A low-power programmable neural spike detection channel with embedded calibration and data compression,"IEEE Transaction on Biomedical Circuits and Systems, vol. 6, no. 4, pp. 87–100, 2012.
36. S. Song, M. J. Rooijakkers, P. Harpe et al., "A 430 nW 64 nV/Hzcurrent-reuse telescopic amplifier for neural recording applications," inProceedings of the IEEE Biomedical Circuits and Systems Conference (BiOCAS '13), pp. 322–325, Rotterdam, The Netherlands, October-November 2013.
37. F. Zhang, J. Holleman, and B. P. Otis, "Design of ultra-low power biopotential amplifiers for biosignal acquisition applications,"IEEE Transactions on Biomedical Circuits and Systems, vol. 6, no. 4, pp. 344–355, 2012.

38. S. Song, M. J. Rooijakkers, P. Harpe et al., “A 430nW 64nV/√Hz current-reuse telescopic amplifier for neural recording applications,” inProceedings of the IEEE Biomedical Circuits and Systems Conference (BiOCAS '13), pp. 322–325, Rotterdam, The Netherlands, October-November 2013.
39. X. Zou, L. Liu, J. H. Cheong et al., “A 100-channel 1-mW implantable neural recording IC,”IEEE Transactions on Circuits and Systems I: Regular Papers, vol. 60, no. 10, pp. 2584–2596, 2013.
40. P. Kmon and P. Gryboś, “Energy efficient low-noise multichannel amplifier in submicron CMOS process,”IEEE Transactions on Circuits and Systems I: Regular Papers, vol. 60, no. 7, pp. 1764–1775, 2013.
41. J. Gak, M. R. Miguez, and A. Arnaud, “Nanopower OTAs with improved linearity and low input offset using bulk degeneration,”IEEE Transactions on Circuits and Systems I: Regular Papers, vol. 61, no. 3, pp. 689–698, 2014.
42. V. Majidzadeh, A. Schmid, and Y. Leblebici, “Energy efficient low-noise neural recording amplifier with enhanced noise efficiency factor,”IEEE Transactions on Biomedical Circuits and Systems, vol. 5, no. 3, pp. 262–271, 2011.
43. M. S. J. Steyaert and W. M. C. Sansen, “A micropower low-noise monolithic instrumentation amplifier for medical purposes,”IEEE Journal of Solid-State Circuits, vol. 22, no. 6, pp. 1163–1168, 1987.
44. K. A. Ng and Y. P. Xu, “A compact, low input capacitance neural recording amplifier,”IEEE Transaction on Biomedical Circuits and Systems, vol. 7, no. 5, pp. 610–620, 2013.
45. B. Gosselin, M. Sawan, and C. A. Chapman, “A low-power integrated bioamplifier with active low-frequency suppression,”IEEE Transactions on Biomedical Circuits and Systems, vol. 1, no. 3, pp. 184–192, 2007.
46. C. Y. Wu, W. M. Chen, and L. T. Kuo, “A CMOS power-efficient low-noise current-mode front-end amplifier for neural signal recording,”IEEE Transaction on Biomedical Circuits and Systems, vol. 7, no. 2, pp. 107–114, 2013.
47. Demosthenous and I. F. Triantis, “An adaptive ENG amplifier for tripolar cuff electrodes,”IEEE Journal of Solid-State Circuits, vol. 40, no. 2, pp. 412–420, 2005.
48. R. Rieger, M. Schuettler, D. Pal et al., “Very low-noise ENG amplifier system using CMOS technology,”IEEE Transactions on Neural Systems and Rehabilitation Engineering, vol. 14, no. 6, pp. 427–437, 2006.
49. Y. J. Jung, B. S. Park, H. M. Kwon et al., “A novel BJT structure implemented using CMOS processes for high-performance analog circuit applications,”IEEE Transactions on Semiconductor Manufacturing, vol. 25, no. 4, pp. 549–554, 2012.

50. M. Degrauwe, E. Vittoz, and I. Verbauwhede, "A micropower CMOS instrumentation amplifier,"IEEE Journal of Solid-State Circuits, vol. 20, no. 3, pp. 805–807, 1985.
51. Worapishet, A. Demosthenous, and X. Liu, "A CMOS instrumentation amplifier with 90-dB CMRR at 2-MHz using capacitive neutralization: analysis, design considerations, and implementation,"IEEE Transactions on Circuits and Systems I: Regular Papers, vol. 58, no. 4, pp. 699–710, 2011.
52. Demosthenous, I. Pachnis, D. Jiang, and N. Donaldson, "An integrated amplifier with passive neutralization of myoelectric interference from neural recording tripoles,"IEEE Sensors Journal, vol. 13, no. 9, pp. 3236–3248, 2013.
53. Rodríguez-Pérez, M. Delgado-Restituto, and F. Medeiro, "A 515 nW, 0–18 dB programmable gain analog-to-digital converter for in-channel neural recording interfaces,"IEEE Transactions on Biomedical Circuits and Systems, 2013.
54. Y. Ming, D. A. Borton, J. Aceros, W. R. Patterson, and A. V. Nurmikko, "A 100-channel hermetically sealed implantable device for chronic wireless neurosensing applications,"IEEE Transactions on Biomedical Circuits and Systems, vol. 7, no. 2, pp. 115–128, 2013.
55. M. S. Chae, Z. Yang, M. R. Yuce, L. Hoang, and W. Liu, "A 128-channel 6 mW wireless neural recording IC with spike feature extraction and UWB transmitter,"IEEE Transactions on Neural Systems and Rehabilitation Engineering, vol. 17, no. 4, pp. 312–321, 2009.
56. H. Gao, R. M. Walker, P. Nuyujukian et al., "HermesE: a 96-channel full data rate direct neural interface in 0.13 μm CMOS,"IEEE Journal of Solid-State Circuits, vol. 47, no. 4, pp. 1043–1055, 2012.
57. S. Gibson, J. W. Judy, and D. Marković, "Spike sorting: the first step in decoding the brain: the first step in decoding the brain,"IEEE Signal Processing Magazine, vol. 29, no. 1, pp. 124–143, 2012.
58. R. R. Harrison, R. J. Kier, C. A. Chestek et al., "Wireless neural recording with single low-power integrated circuit,"IEEE Transactions on Neural Systems and Rehabilitation Engineering, vol. 17, no. 4, pp. 322–329, 2009.
59. M. Sodagar, G. E. Perlin, Y. Yao, K. Najafi, and K. D. Wise, "An implantable 64-channel wireless microsystem for single-unit neural recording,"IEEE Journal of Solid-State Circuits, vol. 44, no. 9, pp. 2591–2604, 2009.
60. S. Gibson, J. W. Judy, and D. Marković, "Technology-aware algorithm design for neural spike detection, feature extraction, and dimensionality reduction,"IEEE Transactions on Neural Systems and Rehabilitation Engineering, vol. 18, no. 5, pp. 469–478, 2010.

61. V. Karkare, S. Gibson, and D. Marković, “A 130-μ W, 64-channel neural spike-sorting DSP chip,”IEEE Journal of Solid-State Circuits, vol. 46, no. 5, pp. 1214–1222, 2011.
62. T. M. Seese, H. Harasaki, G. M. Saidel, and C. R. Davies, “Characterization of tissue morphology, angiogenesis, and temperature in the adaptive response of muscle tissue to chronic heating,”Laboratory Investigation, vol. 78, no. 12, pp. 1553–1562, 1998.
63. K. G. Oweiss, A. Mason, Y. Suhail, A. M. Kamboh, and K. E. Thomson, “A scalable wavelet transform VLSI architecture for real-time signal processing in high-density intra-cortical implants,”IEEE Transactions on Circuits and Systems I: Regular Papers, vol. 54, no. 6, pp. 1266–1278, 2007.
64. D. L. Donoho, “Compressed sensing,”IEEE Transactions on Information Theory, vol. 52, no. 4, pp. 1289–1306, 2006.
65. Z. Charbiwala, V. Karkare, S. Gibson, D. Marković, and M. B. Srivastava, “Compressive sensing of neural action potentials using a learned union of supports,” inProceedings of the 8th International Conference on Body Sensor Networks (BSN '11), pp. 53–58, Dallas, Tex, USA, May 2011.
66. Y. Suo, J. Zhang, R. Etienne-Cummings, T. D. Tran, and S. Chin, “Energy-efficient two-stage compressed sensing method for implantable neural recordings,” inProceedings of the IEEE Biomedical Circuits and Systems Conference (BiOCAS '13), pp. 150–153, Rotterdam, The Netherlands, October-November 2013.
67. Rodríguez-Pérez, J. Masuch, J. A. Rodríguez-Rodríguez, M. Delgado-Restituto, and A. Rodríguez-Vázquez, “A 64-channel inductively-powered neural recording sensor array,” inProceedings of the IEEE Biomedical Circuits and Systems Conference (BiOCAS '12), pp. 228–231, Hsinchu, Taiwan, November 2012.
68. J. Simpson and M. Ghovanloo, “An experimental study of voltage, current, and charge controlled stimulation front-end circuitry,” inProceedings of the IEEE International Symposium on Circuits and Systems (ISCAS '07), pp. 325–328, New Orleans, La, USA, May 2007.
69. X. Liu, A. Demosthenous, and N. Donaldson, “An integrated implantable stimulator that is fail-safe without off-chip blocking-capacitors,”IEEE Transactions on Biomedical Circuits and Systems, vol. 2, no. 3, pp. 231–244, 2008.
70. P. J. Langlois, A. Demosthenous, I. Pachnis, and N. Donaldson, “High-power integrated stimulator output stages with floating discharge over a wide voltage range for nerve stimulation,”IEEE Transactions on Biomedical Circuits and Systems, vol. 4, no. 1, pp. 39–48, 2010.
71. D. Jiang, A. Demosthenous, T. Perkins, X. Liu, and N. Donaldson, “A stimulator ASIC featuring versatile management for vestibular

prostheses,"IEEE Transactions on Biomedical Circuits and Systems, vol. 5, no. 2, pp. 147–159, 2011.

72. X. Liu, A. Demosthenous, and N. Donaldson, "An integrated stimulator with DC-isolation and fine current control for implanted nerve tripoles,"IEEE Journal of Solid-State Circuits, vol. 46, no. 7, pp. 1701–1714, 2011.
73. X. Liu, A. Demosthenous, A. Vanhoestenberghe, D. Jiang, and N. Donaldson, "Active Books: the design of an implantable stimulator that minimizes cable count using integrated circuits very close to electrodes,"IEEE Transactions on Biomedical Circuits and Systems, vol. 6, no. 3, pp. 216–227, 2012.
74. S. K. Kelly and J. L. Wyatt Jr., "A power-efficient voltage-based neural tissue stimulator with energy recovery," inProceedings of the IEEE Solid-State Circuits Conference (ISSCC '04), San Francisco, Calif, USA, February 2004.
75. S. K. Kelly and J. L. Wyatt Jr., "A power-efficient neural tissue stimulator with energy recovery,"IEEE Transactions on Biomedical Circuits and Systems, vol. 5, no. 1, pp. 20–29, 2011.
76. X. Liu, A. Demosthenous, and N. Donaldson, "Implantable stimulator failures: causes, outcomes, and solutions," inProceedings of the 29th Annual International Conference of the IEEE Engineering in Medicine and Biology Society (EMBC '07), pp. 5786–5789, Lyon, France, August 2007.
77. Nonclercq, L. Lonys, A. Vanhoestenberghe, A. Demosthenous, and N. Donaldson, "Safety of multi-channel stimulation implants: a single blocking capacitor per channel is not sufficient after single-fault failure,"Medical & Biological Engineering & Computing, vol. 50, no. 4, pp. 403–410, 2012.
78. J.-J. Sit and R. Sarpeshkar, "A low-power blocking-capacitor-free charge-balanced electrode-stimulator chip with lesst than 6 nA DC error for 1-mA: full-scale stimulation,"IEEE Transactions on Biomedical Circuits and Systems, vol. 1, no. 3, pp. 172–183, 2007.
79. M. Ortmanns, A. Rocke, M. Gehrke, and H.-J. Tiedtke, "A 232-channel epiretinal stimulator ASIC,"IEEE Journal of Solid-State Circuits, vol. 42, no. 12, pp. 2946–2959, 2007.
80. Williams and T. G. Constandinou, "An energy-efficient, dynamic voltage scaling neural stimulator for a proprioceptive prosthesis,"IEEE Transactions on Biomedical Circuits and Systems, vol. 7, no. 2, pp. 129–139, 2013.
81. S. K. Arfin and R. Sarpeshkar, "An energy-efficient, adiabatic electrode stimulator with inductive energy recycling and feedback current regulation,"IEEE Transactions on Biomedical Circuits and Systems, vol. 6, no. 1, pp. 1–14, 2012.
82. U. Çilingiroğlu and S. İpek, "A zero-voltage switching technique for minimizing the current-source power of implanted stimulators,"IEEE Transactions on Biomedical Circuits and Systems, vol. 7, no. 4, pp. 469–479, 2013.

83. D. Jiang, A. Demosthenous, T. Perkins, D. Cirmirakis, X. Liu, and N. Donaldson, "An implantable 3-D vestibular stimulator with neural recording," inProceedings of the 38th European Solid-State Circuits Conference (ESSCIRC '12), pp. 277–280, Bordeaux, France, September 2012.
84. P. Bradley, "Wireless medical implant technology—recent advances and future developments," inProceedings of the 37th European Solid-State Circuits Conference (ESSCIRC '11), pp. 54–58, Helsinki, Finland, September 2011.
85. H. Yu, C.-M. Tang, and R. Bashirullah, "An asymmetric RF tagging IC for ingestible medication compliance capsules," inProceedings of the IEEE Radio Frequency Integrated Circuits Symposium (RFIC '09), pp. 101–104, Boston, Mass, USA, June 2009.
86. D. Cirmirakis, D. Jiang, A. Demosthenous, N. Donaldson, and T. Perkins, "A telemetry operated vestibular prosthesis," inProceedings of the 19th International Conference on Electronics, Circuits, and Systems (ICECS '12), pp. 576–578, Seville, Spain, December 2012.
87. M. Sawan, Y. Hu, and J. Coulombe, "Wireless smart implants dedicated to multichannel monitoring and microstimulation,"IEEE Circuits and Systems Magazine, vol. 5, no. 1, pp. 21–39, 2005.
88. M. A. Hannan, S. M. Abbas, S. A. Samad, and A. Hussain, "Modulation techniques for biomedical implanted devices and their challenges,"Sensors, vol. 12, no. 1, pp. 297–319, 2012.
89. M. Silay, C. Dehollain, and M. Declercq, "Inductive power link for a wireless cortical implant with biocompatible packaging," inProceedings of the 9th IEEE Sensors Conference (SENSORS '10), pp. 94–98, Kona, Hawaii, USA, November 2010.
90. M. A. Adeeb, A. B. Islam, M. R. Haider, F. S. Tulip, M. N. Ericson, and S. K. Islam, "An inductive link-based wireless power transfer system for biomedical applications,"Active and Passive Electronic Components, vol. 2012, Article ID 879294, 11 pages, 2012.
91. R. R. Harrison, "Designing efficient inductive power links for implantable devices," inProceedings of the IEEE International Symposium on Circuits and Systems (ISCAS '07), pp. 2080–2083, New Orleans, La, USA, May 2007.
92. F. E. Terman,Electronic and Radio Engineering, McGraw-Hill, New York, NY, USA, 4th edition, 1955.
93. N. Donaldson and T. A. Perkins, "Analysis of resonant coupled coils in the design of radio frequency transcutaneous links,"Medical & Biological Engineering & Computing, vol. 21, no. 5, pp. 612–627, 1983.
94. D. Cirmirakis,Novel telemetry system for closed loop vestibular prosthesis [Ph.D. thesis], University College London, London, UK, 2013.

95. D. C. Galbraith, M. Soma, and R. L. White, "A wide-band efficient inductive transdermal power and data link with coupling insensitive gain,"IEEE Transactions on Biomedical Engineering, vol. 34, no. 4, pp. 265–275, 1987.
96. G. Wang, W. Liu, M. Sivaprakasam, and G. A. Kendir, "Design and analysis of an adaptive transcutaneous power telemetry for biomedical implants,"IEEE Transactions on Circuits and Systems I: Regular Papers, vol. 52, no. 10, pp. 2109–2117, 2005.
97. P. Si, A. P. Hu, S. Malpas, and D. Budgett, "A frequency control method for regulating wireless power to implantable devices,"IEEE Transactions on Biomedical Circuits and Systems, vol. 2, no. 1, pp. 22–29, 2008.
98. G. Simard, M. Sawan, and D. Massicotte, "High-speed OQPSK and efficient power transfer through inductive link for biomedical implants,"IEEE Transactions on Biomedical Circuits and Systems, vol. 4, no. 3, pp. 192–200, 2010.
99. N. Donaldson, "Passive signalling via inductive coupling,"Medical & Biological Engineering & Computing, vol. 24, no. 2, pp. 223–224, 1986.
100. Z. Tang, B. Smith, J. H. Schild, and P. H. Peckham, "Data transmission from an implantable biotelemeter by load-shift keying using circuit configuration modulator,"IEEE Transactions on Biomedical Engineering, vol. 42, no. 5, pp. 524–528, 1995.
101. W. Xu, Z. Luo, and S. Sonkusale, "Fully digital BPSK demodulator and multilevel LSK back telemetry for biomedical implant transceivers,"IEEE Transactions on Circuits and Systems II: Express Briefs, vol. 56, no. 9, pp. 714–718, 2009.
102. Zhou and N. Donaldson, "A fast passive data transmission method for ENG telemetry,"Neuromodulation, vol. 6, no. 2, pp. 116–121, 2003.
103. D. Cirmirakis, D. Jiang, A. Demosthenous, N. Donaldson, and T. Perkins, "A fast passive phase shift modulator for inductively coupled implanted medical devices," inProceedings of the 38th European Solid-State Circuits Conference (ESSCIRC '12), pp. 301–304, Bordeaux, France, September 2012.
104. Kiani and M. Ghovanloo, "A 20-Mb/s pulse harmonic modulation transceiver for wideband near-field data transmission,"IEEE Transactions on Circuits and Systems II: Express Briefs, vol. 60, no. 7, pp. 382–386, 2013.
105. Ghovanloo and K. Najafi, "A wideband frequency-shift keying wireless link for inductively powered biomedical implants,"IEEE Transactions on Circuits and Systems I: Regular Papers, vol. 51, no. 12, pp. 2374–2383, 2004.
106. Z. Luo and S. Sonkusale, "A novel BPSK demodulator for biological implants,"IEEE Transactions on Circuits and Systems I: Regular Papers, vol. 55, no. 6, pp. 1478–1484, 2008.
107. P. P. Mercier, A. C. Lysaght, S. Bandyopadhyay, A. P. Chandrakasan, and K. M. Stankovic, "Energy extraction from the biologic battery in the inner ear,"Nature Biotechnology, vol. 30, no. 12, pp. 1240–1243, 2012.

108.Saeidi, M. Schuettler, A. Demosthenous, and N. Donaldson, "Technology for integrated circuit micropackages for neural interfaces, based on gold-silicon wafer bonding,"Journal of Micromechanics and Microengineering, vol. 23, Article ID 075021, 12 pages, 2013.

109.Mohseni and K. Najafi, "A fully integrated neural recording amplifier with DC input stabilization,"IEEE Transactions on Biomedical Engineering, vol. 51, no. 5, pp. 832–837, 2004.

110.W. S. Liew, X. D. Zou, L. B. Yao, and Y. Lian, "A 1-V 60μW 16-channel interface chip for implantable neural recording," inProceedings of the 31st IEEE Custom Integrated Circuits Conference (CICC '09), pp. 507–510, San Jose, Calif, USA, September 2009.

111.S. Rai, J. Holleman, J. N. Pandey, F. Zhang, and B. Otis, "A 500μW neural tag with 2μVrmsAFE and frequency-multiplying MICS/ISM FSK transmitter," inProceedings of the IEEE International Solid-State Circuits Conference (ISSCC '09), vol. 1, pp. 212–213, San Francisco, Calif, USA, February 2009.

112.F. Shahrokhi, K. Abdelhalim, D. Serletis, P. L. Carlen, and R. Genov, "The 128-channel fully differential digital integrated neural recording and stimulation interface,"IEEE Transactions on Biomedical Circuits and Systems, vol. 4, no. 3, pp. 149–161, 2010.

113.R. R. Harrison, P. T. Watkins, R. J. Kier et al., "A low-power integrated circuit for a wireless 100-electrode neural recording system,"IEEE Journal of Solid-State Circuits, vol. 42, no. 1, pp. 123–133, 2007.

114.Bonfanti, M. Ceravolo, G. Zambra et al., "A multi-channel low-power IC for neural spike recording with data compression and narrowband 400-MHz MC-FSK wireless transmission," inProceedings of the 36th European Solid-State Circuits Conference (ESSCIRC '10), pp. 330–333, Seville, Spain, September 2010.

115.S. B. Lee, H.-M. Lee, M. Kiani, U.-M. Jow, and M. Ghovanloo, "An inductively powered scalable 32-channel wireless neural recording system-on-a-chip for neuroscience applications,"IEEE Transactions on Biomedical Circuits and Systems, vol. 4, no. 6, pp. 360–371, 2010.

116.K. Abdelhalim, L. Kokarovtseva, J. L. P. Velazquez, and R. Genov, "915-MHz FSK/OOK wireless neural recording SoC with 64 mixed-signal FIR filters,"IEEE Journal of Solid-State Circuits, vol. 48, no. 10, pp. 2478–2493, 2013.

CITATION

Andreas Demosthenous, "Advances in Microelectronics for Implantable Medical Devices,"Advances in Electronics, vol. 2014, Article ID 981295, 21 pages, 2014. doi:10.1155/2014/981295.

CHAPTER 3

Applications of Fianite in Electronics

Alexander N. Buzynin,[1] *Yury N. Buzynin,*[2] *and Vitaly A. Panov*[1]

[1]A. M. Prokhorov General Physics Institute, Russian Academy of Sciences, Moscow 119991, Russia
[2]Institute for Physics of Microstructures, Russian Academy of Sciences, Nizhny Novgorod 603950, Russia

ABSTRACT

Fianite or yttrium stabilized zirconia (YSZ) solid solutions single crystals were known worldwide as jewelry material. The review is devoted to novel applications of the material in the field of microelectronics. A number of modern aspects of the application of fianite in micro-, opto- and SHF-electronics were analyzed in this paper. It was demonstrated that fianite is an extremely promising multipurpose material for new electronic technologies due to unique combination of physical and chemical properties. As a substrate and buffer layer for the epitaxy of Si, Ge, GeSi and $A^{III}B^{V}$ compounds (GaAs, InGaAs, GaSb, InAs, GaN, AlN), fianite has a number of advantages over the other dielectric materials. The use of fianite (as well as ZrO_2 and HfO_2 oxides) instead of SiO_2 as gate dielectrics in CMOC technology seems to be of peculiar interest. The unique properties of fianite as protecting, stabilizing and antireflecting coatings in electronics and optoelectronic devices have been outlined. A comparative study of the performance characteristics of fianite and conventional materials has been carried out.

INTRODUCTION

The further progress in electronics is connected with application of new materials. Fianite is a material of such a kind. Industrial technology of synthesis of fianite has been for the first time developed in Russia in the Lebedev Physical Institute of the Russian Academy of Sciences (FIAN in Russian), so the crystals were entitled after the Institute [1]. Serial production of the crystals has been already started in the early seventies of the twentieth century. Currently, fianite crystals are in the second position by the volume of worldwide production following silicon [2, 3]. Fianite single crystals, -zirconia-based solid solutions (or "yttrium stabilized zirconia" YSZ), were known as jewelry stone imitation materials. Recently, in the countries with the developed microelectronics a significant growth of interest to various aspects of fianite application in semiconductor technologies has been observed. Fianite is an extremely promising multipurpose material for new optoelectronics technologies due to its unique combination of physical and chemical properties. It can be used in, virtually, all of the main technological stages of the production of electronic devices (Figure 1).

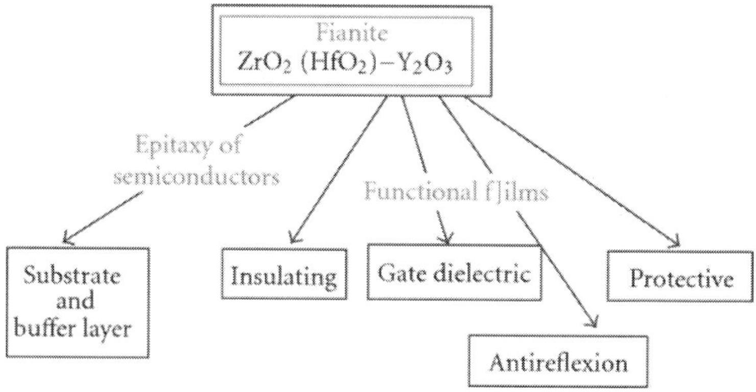

Figure 1. Application of fianite in electronics.

A number of application prospects of fianite in modern electronics are considered in this paper.

FIANITE AS A SUBSTRATE AND BUFFER LAYER FOR EPITAXY OF SEMICONDUCTORS, MULTILAYER HETEROSTRUCTURES, AND SUPERLATTICES

Appropriate conditions of growth of mirror-flat single-crystalline films of : Ge, GeSi, and $A^{III}B^{V}$ compounds: GaAs, InGaAs, GaSb, InAs, GaN, AlN, and InN as well as multilayer InGaAlAs heterostructures and GaSb/InAs supperlattices on fianite substrates, as well as on Si and GaAs substrates coated with fianite buffer layer, using MOCVD, HW-CVD, and laser deposition techniques have been elaborated. All of these films have been for the first time synthesized in Russia.

A Fianite as a Substrate and Buffer Layer for Si, Si-Ge, and AIIIBV Compounds Epitaxy

Fianite has a number of advantages as a substrate and buffer layer at Si and $A^{III}B^{V}$ compounds epitaxy, as compared with other dielectric materials [4–13].

In comparison with the other dielectrics, there are the following merits of fianite in application as a substrate material and buffer layer for Si and $A^{III}B^{V}$ compounds epitaxy:

i. High resistivity—>10^{12} Ohm.cm at 300K.
ii. Similarly to Si, Ge, and $A^{III}B^{V}$ compounds, it is of cubic structure (in contrast to hexagonal of sapphire).
iii. It is possible to alter fianite cubic lattice constant in solid solutions by varying the ratio of the main (zirconium or hafnium dioxide) and stabilizing oxides (yttria, rare earth oxides from gadolinium to lutetium, and alkaline-earth oxides) that allows an optimum matching between substrate and cubic lattice of semiconductor films thus improving its structural perfection.
iv. Negligible value of diffusion coefficients of cations up to 1000–1200°C temperature range that excludes interdiffusion of impurities between the substrate and film and prevents undesirable doping (i.e., typical for sapphire), which can damage heteroepitaxial layers through penetration of impurity atoms.
v. Due to its excellent stability at elevated temperatures, the upper limit of the corresponding structure operational temperatures depends on physical properties of a semiconductor only. Elevated temperature is not critical for the substrate.
vi. Broad spectral range of transmission (260–7500nm) completely covers actual absorbance and emission of Si, $A^{III}B^{V}$ compounds and its solid solutions. That makes "semiconductor-on-fianite" structures very promising for the development of various optoelectronic devices with improved operational parameters (avalanche photodiodes, light-emitting and laser diodes, etc.).
vii. Application of thin layers of fianite on Si and GaAs instead of its monolithic substrates allows avoiding spatial limitations of the

structures and decreasing the net cost. At the same time, the structures on “fianite/Si” and “fianite/GaAs” episubstrates have better heat conductivity in comparison with the structures on monolithic substrates.

The first epitaxial Si films on YSZ were grown in [6]. The first successful results on epitaxial MOCVD growth of various $A^{III}B^{V}$ compounds (GaAs, InAs, InGaAs, AlGaAs, GaAsN, and GaN) on YSZ are presented in a number of studies [6, 11, 12], InN on YSZ in [14, 15]. In [13, 16] a capillary epitaxy technique—the new effective way of heteroepitaxy was developed. It has been shown that the use of capillary forces in the method positively influences both on the mechanism of epitaxial growth and on quality of $A_{III}B_{V}$ epitaxial films and also reduces the minimum thickness of a continuous layer [16, 17].

An application of fianite as either monolithic substrate or buffer layer in “semiconductor-on-dielectric” technology is of peculiar importance for micro- and optoelectronics. The technology allows improving such characteristics of integrated circuits as operation speed, critical operational temperature, and radiation resistance.

Due to a decrease of the loss of current and stray capacitance, energy consumption of the devices is decreasing. Moreover, the devices based on “semiconductor-on-dielectric” structures are more reliable, especially under extreme operational conditions. Currently, “silicon-on-insulator” structures are one of the most dynamically developing directions in the field of semiconductor materials science. However, electrophysical and operational parameters of the devices as well as its radiation resistance and reliability significantly suffer because of structure imperfection of silicon layers. In case of “silicon-on-sapphire” structures, the imperfection is determined, in particular, by a difference in crystallographic structure of silicon and sapphire, as well as by autodoping of a silicon film by aluminum penetrating from the sapphire substrate in concentrations up to 10^{18}–10^{20}cm^{-3}. Considering crystal-chemical and physical characteristics of fianite, the material is more preferential for the epitaxy of Si as an alternative substrate in comparison with sapphire.

Silicon-on-Fianite Epitaxial Structures

The first studies on silicon epitaxy on fianite single-crystal substrates have been carried out in France and USA [6, 7]. Silicon films on fianite substrate were deposited by chloride and hydride epitaxy at 900–1100°C. The films obtained were of polycrystalline structure and, consequently, of rather poor electrophysical parameters. However, at the same time, it was shown that silicon-on-fianite structures sustaining actually all advantages of silicon-on-sapphire are free from its principal drawbacks.

At the epitaxy of Si on fianite, a formation of SiO_2 intermediate layer between the film and the substrate was observed [7, 8]. Subsequent annealing of the structure led to the increase of SiO_2 layer thickness. It was demonstrated [8] that the layer can improve properties of silicon-on-fianite epitaxial structure because its formation:

i. Removes mechanical stress in the layer-substrate interface;
ii. Smoothens over negative effect occurring due to a difference of linear expansion coefficients between fianite and silicon;
iii. Improves insulation of the integrated circuit elements (ICE) based on Si;
iv. Acts as a barrier for metal impurities diffusing from the substrate and forming deep levels in silicon.

The formation of SiO_2 intermediate layer at high-temperature epitaxy is associated with peculiar properties of fianite. In contrast to the other dielectrics, fianite features with a unique peculiarity as a solid electrolyte: starting from 650°C, it becomes actually oxygen-transparent due to high mobility of oxygen. The reason for significant mobility of oxygen in fianite crystals is an occurrence of oxygen vacancies due to Zr^{+4} to Y^{+3} cation substitution at formation of the solid solution. High mobility of oxygen in fianite crystals is determined by an occurrence of oxygen vacancies at $ZrO_2(HfO_2)$–R_2O_3 (here: R-Y, Gd–Yb) solid solutions formation due to $Zr^{+4}(Hf^{+4})$ to R^{+3}cation substitution. The process results in oxygen nonstoichiometricZ_RO_2 (HFO_2)based phase [4]. Because of the high mobility of oxygen at high temperature of the epitaxy (900–1000°C) used in [6–8], the formation of ether SiO_2 continuous layer or its islets between the substrate and the film was shown to be inevitable.

The phenomenon occurs even at the epitaxy initial stages when a continuous epitaxial film is forming. It was shown [9] that the formation of SiO_2 layer or isles at the initial stage of molecular-beam epitaxy on fianite results in 3-dimensional mechanism of growth, formation of structural defects and hindered the synthesis of Si films of single-crystal structure. The occurrence of the isles at the initial epitaxy stages and the polycentric growth of Si layers were shown possible to avoid only by using a set of techniques, those which prevent diffusion of oxygen from the substrate to the film at the initial stage of the process. In particular, high structural perfection of the Si-on-fianite films was achieved by using a low-temperature ($T < 650°C$) molecular-beam epitaxy [7].

Ge and GeSi Films on Fianite Substrates

Growth of Ge and Ge-Si heterostructures on fianite substrates was carried out using HWCVD installation. Base pressure in the chamber ~110^{-8}torr was maintained by pumping-down using two heteroionic pumps. A high-vacuum gate was used for isolation of the growth cell and the pumps from other parts of the vacuum system. Forepumping of the chamber was performed using a diffusion pump. The diffusion pump allowed to exhaust any gas (including GeH_4) both in atomic and molecular state. FM-1 oil with low vapor pressure was used as a pressure fluid. There was a nitrogen trap above the diffusion pump preventing reverse diffusion of the oil from preevacuation and diffusion pumps into the growth cell. The (100) and (111) oriented fianite single-crystal plates were used as substrates. Silicon atomic beam was maintained by sublimation of the element single-crystal (high resistance) in form of 4 × 4 × 90mm ingot sections. The sources were mounted on the cooled current leads. There was a Ta plate of 80 × 5 × 0.5mm size installed in one of the sources position.

Before the epitaxial growth, the sources and substrates were subjected to 10min annealing at 1350 and 1250°C, respectively, then temperature of the source was increased to 1380°C, as the substrate temperature was decreased to assigned values (600–700°C) and the buffer layer was grown. The pressure in the cell corresponded to basic one.

In order to grow Ge layers, the cell was filled with GeH_4 up to 110^{-3}–510^{-6}torr and the pressure was maintained constant by a system of the gas feeding. Simultaneously, the Ta plate situated in vicinity of the substrate was heated to T=200^{0}C. With the purpose to avoid destruction of germane on evaporators (Ti) following preepitaxial annealing of the sources and substrates, the sublimating pumps were switched off and the growth was carried out at pumping down using only diffusion- and booster-pumps. It is worth to note that the gas filling up to such high pressure (~10^{-3}torr) is impossible in MBE installations with electron-beam heating. Germane pressure in the cell was tentatively assigned by ionization vacuum gage indications. Nevertheless, this peculiarity in GeH_4 pressure measurement did not impede the controlled growth of Ge films at 700–750°C temperature of the substrate. The films were continuous and homogeneous. GeSi solid solutions with up to 80% Si content were also obtained on (111) and (100) fianite substrates. Vacuum annealing at 1250°C during 10min was used as a preepitaxy treatment. The growth was carried out under 510^{-4}torr germane pressure and at 600°C substrate temperature. Simultaneously, the Ta plate positioned in vicinity of the substrate was heated to 1200°C. The heteroepitaxial Ge films obtained show high structural perfection. X-ray rocking curve (XRC) FWHM values were 0.31° for Ge film. The surface morphology of the Ge epitaxial layers grown on (100) and (111) fianite substrates as well as the peaks of Raman scattering near 300cm^{-1} are identical to those of bulk Ge. Therefore, it is possible to conclude that there are no stains in the Ge/fianite layer.

Epitaxial Films of A^{III}B^{V} on Fianite

Crystallochemical and physical properties of fianite are favorable not only for silicon but also for compounds epitaxy (Table 1).

Table 1. Some properties of fianite crystals and A^{III}B^{V} compounds.

Crystal	Lattice Type	a, Å	T_m, °C (melting point)	Thermal expansion coefficients 10^{-6} deg^{-1}	E_g, eV
$(ZrO_2)_{100-x}(Y_2O_3)_x$	Cubic (fluorite)	5.141 (x = 10) 5.157 (x = 15) 5.198 (x = 21)	2800	11.4 (15–1000°C)	
GaAs	Cubic (sphalerite)	5.65	1283	5.4	1.43
GaP	Cubic (sphalerite)	5.445	1467	4.7	2.26
GaN	Hexagonal (wurtzite)	a = 3.186; c = 5.178	1700	5.6; 7.8	3.4
GaN	Cubic (sphalerite)	4.52	1700	3.9	3.2
InN	Hexagonal (wurtzite)	a = 3.54 c = 5.70	1200	12.7	0.7
InN	Cubic (sphalerite)	4.98	1200	4.4	0.67

First successful results on growth of A^{III}B^{V} compound epitaxial films on fianite substrates were presented in [10,18]. GaAs, InAs, GaN, and other A^{III}B^{V} semiconductor compound films have been grown on fianite, as well as on silicon and gallium arsenide with fianite buffer layer substrates by means of metal-organic chemical vapor deposition (MOCVD). A new efficient epitaxy technique-“capillary epitaxy” has been suggested. The technique allowed synthesizing of A^{III}B^{V} compound films by a MOCVD on fianite substrates. Samples of structurally perfect submicron (up to 0.1μ) epitaxial films of A^{III}B^{V} compounds have been obtained using this technique. The samples demonstrated high electrophysical parameters [13, 17–20]. In [21], GaN epitaxial films have been grown on fianite substrates by MOVP technique. It was observed that the epitaxial growth of GaN on fianite significantly depends on conditions of the initial stage of the process.

In [11, 22, 23], fianite substrates were successfully tested for growth of InN heteroepitaxial films. InN films of cubic structure have been grown on

(001) fianite substrates by plasma-stimulated molecular-beam epitaxy (RF-MBE) at 400–490°C temperature. The lattice mismatch of InN and fianite at (001) plane is very low (less than 2.3%), in contrast to 17% for InN sapphire and more than 10% for InN-GaAs. Due to this fact, InN films grown on (001) fianite substrate were superior InN films grown on sapphire [10] and (001) GaAs substrates by its crystallographic perfection [15].

Therefore, fianite is apparently in advance as a substrate for InN epitaxy as compared to sapphire. A new effective method of heteroepitaxy, capillary epitaxy, was proposed in [17]. In particular, this technique allows us to obtain the films of $A^{III}B^{V}$ compounds on fianite using a MOCVD approach.

Deposition of GaAs, GaSb, GaAs:Sb Films, and GaSb/InA Superlattice on Fianite Substrates by Means of Laser Sputtering

Our experiments have shown that the conventional "direct" growth of heteroepitaxial InGaAs films on fianite substrates resulted in the films with rough surface. So the buffer layers were elaborated to improve the results. The buffer layer must have very high structural perfection and mirror-homogeneous surface. A number of experiments were conducted for growth of GaAs, GaSb, and GaAs buffer layers on fianite (100) and (111) substrates as well as GaSb/InAs superlattice by using laser sputtering. This superlattice is working as a filter, which prevents penetration of the defects into InGaAs film and, first of all, formation of growing dislocation. Furthermore, Sb is an effective surfactant which significantly improves the films morphology.

The studies have shown that it was complicated to obtain thin and homogeneous layers of $A^{III}B^{V}$ compounds on fianite substrates. It may be related to rather high mismatching of the lattice parameters of fianite and compounds leading to growth according to the Volmer-Weber mechanism. Formation of the continuous layer occurred through 3-dimensional nuclei, their subsequent growth and joining. Low nuclei density results in the formation of highly inhomogeneous rough surface that hinders subsequent formation of a flat film. A laser sputtering technique is considered to maintain high nuclei density; so, before joining, the nuclei are of sufficiently small size that promotes the formation of a flat continuous film.

Therefore, in order to obtain flat layers, a laser sputtering technique was used in the study.

The Q-switched Nd laser and single-crystal GaAs and InAs targets were used. The superlattices were grown by optical switching of the laser beam between the targets. Mirror-flat GaSb and GaAs:Sb layers as well as penta-periodic InAs/GaSb supperlattices of 0.15μm total thickness were deposited using this technique.

The X-ray diffraction investigations of GaAs:Sb (111) films on fianite (111) showed their single-crystal structure (Figure 2(a)). It was shown that the spectral dependence of photoconductivity of GaSb films on fianite substrates (Figure 2(b)) has a maximum of photoconductivity at the edge of fundamental absorption. This effect may be due to high velocity of the surface recombination.

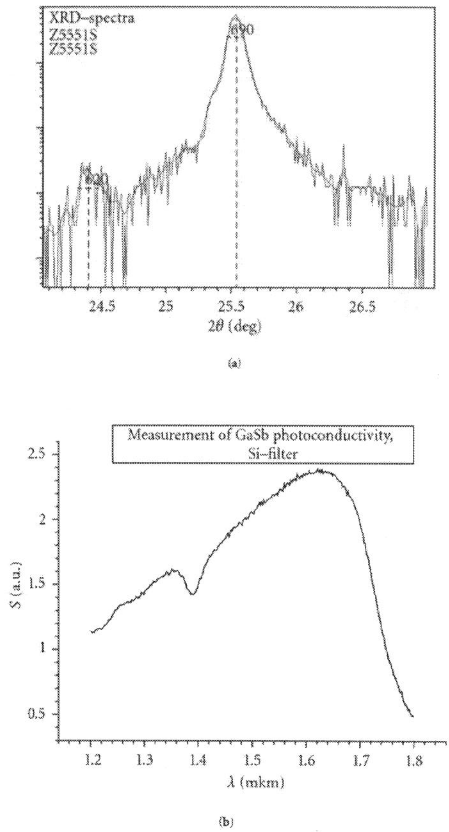

Figure 2. XRD λ(mkm) scan of GaSb (111) film on fianite (111) (a) and photoconductivity of GaSb film on the fianite substrate (b).

The X-ray rocking curve (XRC) FWHM value was 0.23° for GaSb (111) film. The image of the surface of GaAs:Sb (0.2 μm thickness) on fianite is shown in Figure 3(a). It is apparent that the surface of the layer is mirror flat and sufficiently homogeneous. The microrelief of the layer surface is shown in Figure 3(b). According to our estimations, roughness of the layer is less than 4nm (Sq = 0.003778μm).

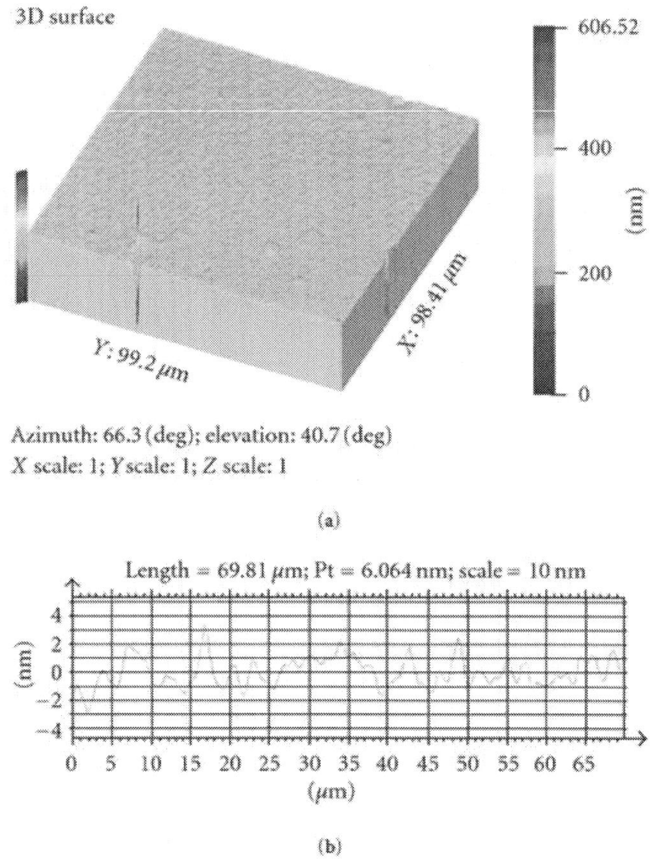

Figure 3. Image of surface (a) and the surface relief (b) of GaAs:Sb buffer on fianite.

In the penta-periodic InAs/GaSb supperlattices of 0.15μm total thickness grown on (111) fianite substrates electron mobility approaches to 580cm^2/V × s. The GaSb layers, as well as InAs/GaSb short-period supperlattices, are suitable for the development of IR detectors operating in a 2-3μm range. In our studies, the structures were used as buffers for A^{III}N growth on fianite substrates.

MOCVD Capillary Epitaxy of III-V Compounds on Fianite

The investigations showed that continuous GaAs films on fianite can be obtained only in a very narrow range of the epitaxial conditions. In particular, a temperature range of 550–600°C is necessary. The minimum thickness of a continuous layer was 1.5–2.0μm. The epitaxial films had polycrystalline structure and rough surface. Structural and electrical properties of GaAs films could be improved using capillary epitaxy. The essence of this method is that a thin (less than 50nm) film of an III-group element is initially deposited on fianite surface and then saturated with a V-group component with the formation of a thin continuous epitaxial III-V layer. Following this procedure, the film growth continues to obtain the necessary thickness under conventional epitaxial conditions.

The use of capillary forces in the first (heteroepitaxial) stage of GaAs film formation led to improvement of epitaxial quality. Electron microscopy of the GaAs films at the initial growth stages showed that the transition from conventional MOCVD growth to capillary epitaxy leads to a change in the growth mechanism. Three-dimensional island mechanism changes to the two-dimensional one with propagation of the growth steps (Figure 4(a)). This process is similar to graphoepitaxy [24, 25] from aqueous solutions with addition of surfactants, where an increase in the substrate wettability also significantly improves the quality of graphoepitaxial layers [24] (Figure 4(b)).

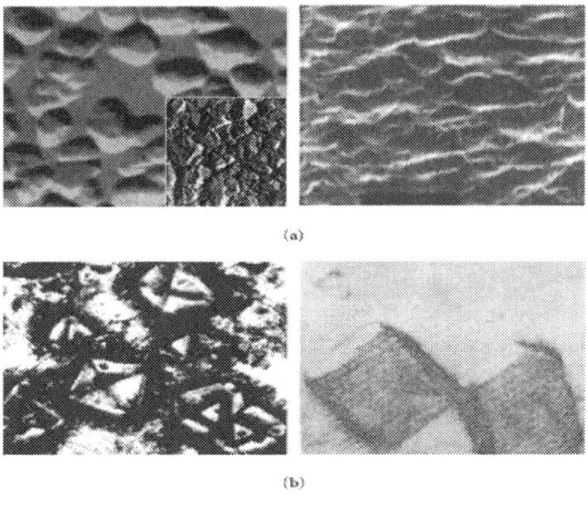

Figure 4. Analogy between the capillary epitaxy and graphoepitaxy. (a) Electron microscopy image of GaAs on YSZ at the initial stage of growth (20000x): conventional MOCVD, height of the islets is up to 3000nm. (b) The capillary epitaxy technique, minimal layer thickness is 50nm, the layer growth is visible [18]. Optical microscopy image of NH_4 J on amorphous Al graphoepitaxy growth: without (c) and with (d) the use of surface-active substances, magnification 100x [24].

In both cases, the height of the crystallization medium (melt or solution) decreases in the initial stage due to the capillary forces. This effect impedes growth of epitaxial nuclei in the direction normal to the substrate surface and facilitates their growth in the tangential direction. As a result, the substrate orienting role increases and a transition to the layer-by-layer growth mechanism occurs with a decrease in the growth step height. Consequently, the minimum height of the continuous film decreases and the film structural quality is improved. It has been shown that the use of capillary forces in this technique improved both the mechanism of epitaxial growth and quality of $A^{III}B^{V}$ epitaxial films. It also reduces the minimum thickness of a continuous layer [14, 18]. Virtually the same approach to deposition of $A^{III}N$ films on various substrates has been successfully applied in studies of the other authors [26].

The use of capillary epitaxy made it possible to decrease minimum thickness of a continuous GaAs/fianite film to 25nm and to improve its structural quality and surface morphology. The technique was also efficient for growing of the other $A^{III}B^{V}$ compounds on fianite.

Deposition of GaAs, AlGaAs, and InGaAs-Based Multilayer Structures on Fianite

The results on epitaxial growth of $A^{III}B^{V}$ compound films obtained in the studies described above were used for obtaining of AlGaAs/InGaAs/GaAs multilayer heterostructures on fianite. These structures were used in FET. Sequential growth $A^{III}B^{V}$ heteroepitaxial layers on fianite substrates was conducted according to the topologic scheme of PHEMT (Pseudomorphic High Electron Mobility Transistor) for microwave frequency FET operating in 10–40GHz range (Table 2) using “Aixtron AIX 200RF” installation. Capillary epitaxy MOCVD technique in 550–600°C temperature range was used.

Table 2. PHEMT heterostructure for FET operating in 10–40GHz range.

n^+ GaAs:Si	$n_{Si} \sim 6 \times 10^{18}$ cm^{-3}	40 nm
i-$Al_xGa_{1-x}As$	$x \sim 0.24$ (>0.23)	25 nm
i-GaAs		~0.6 nm
δ-Si	$n_{Si} \sim 4.5 \times 10^{12}$ cm^{-2}	
i-GaAs		~0.6 nm
i-$Al_xGa_{1-x}As$	$x \sim 0.24$	4 nm
i-GaAs		1 nm
i-$In_yGa_{1-y}As$	$y \sim 0.18$ (<0.2)	11 nm
i-GaAs		30 nm
i-$Al_xGa_{1-x}As$	$x \sim 0.24$	50 nm
i-GaAs	$n < 8 \times 10^{14}$ cm^{-3}	0.5–0.8 μm
CP AlAs/GaAs		(1 nm/2 nm) × 5
GaAs:Sb		100 nm
	Fianite substrate	400 μm

Grown by the "capillary epitaxy" technique series of GaSb and GaAs:Sb buffer layers on (111) and (100) fianite substrates were developed to decrease surface roughness of the PHEMT heterostructure. The buffer layers had a uniform mirror-smooth surface of about 5nm roughness. Application of the developed buffers made it possible to obtain an AlGaAs/InGaAs/GaAs heterostructures with uniform mirror-smooth surface on fianite substrates and to decrease its roughness by a factor of 10 (to 25nm). As a result, sufficiently homogeneous AlGaAs/InGaAs/GaAs multilayer heterostructures with smooth slightly bloom surface were grown on (001) fianite substrates of 50mm diameter. Roughness of the heterostructure surface measured using Talysurf interference microscope (3-dimensional topography) was 0.25μm. This structure was grown using "AIXTRON" installation on (100) fianite ellipsoidal substrate of 2inch major diameter. The surface of the multilayer structure is rather uniform but its roughness reaches the value of 25nm.

Structural perfection of AlGaAs/InGaAs/GaAs multi-layer heterostructures on fianite was investigated by means of XRD. DRON-4 device (Ge(004) monochromator, $CuK\alpha 1$ radiation) was used. $\Theta/2\Theta$ spectra were recorded at symmetric reflection mode by scanning with 0.1 step of the texture maxima rocking. X-ray diffraction spectrum of GaAs (001)/fianite (001) is shown in Figure 5. The peaks of $(Zr,Y)O_2$ (004), $2\theta = 73.4$ substrate and of GaAs (004), $2\theta = 66.05^\circ$ buffer layer were recorded. The width of the layer rocking curve $FWHM_{\omega} = 1'$, that is, the evidence of a mosaic structure of GaAs layer. The grain-boundary angle was ~1°. The use of (111) fianite substrates with GaAs:Sb buffer layers for deposition of AlGaAs/InGaAs/GaAs heterostructures on fianite resulted in the formation of mirror-flat homogeneous surface and 10-fold decrease of the surface roughness (up to 0.025μm).

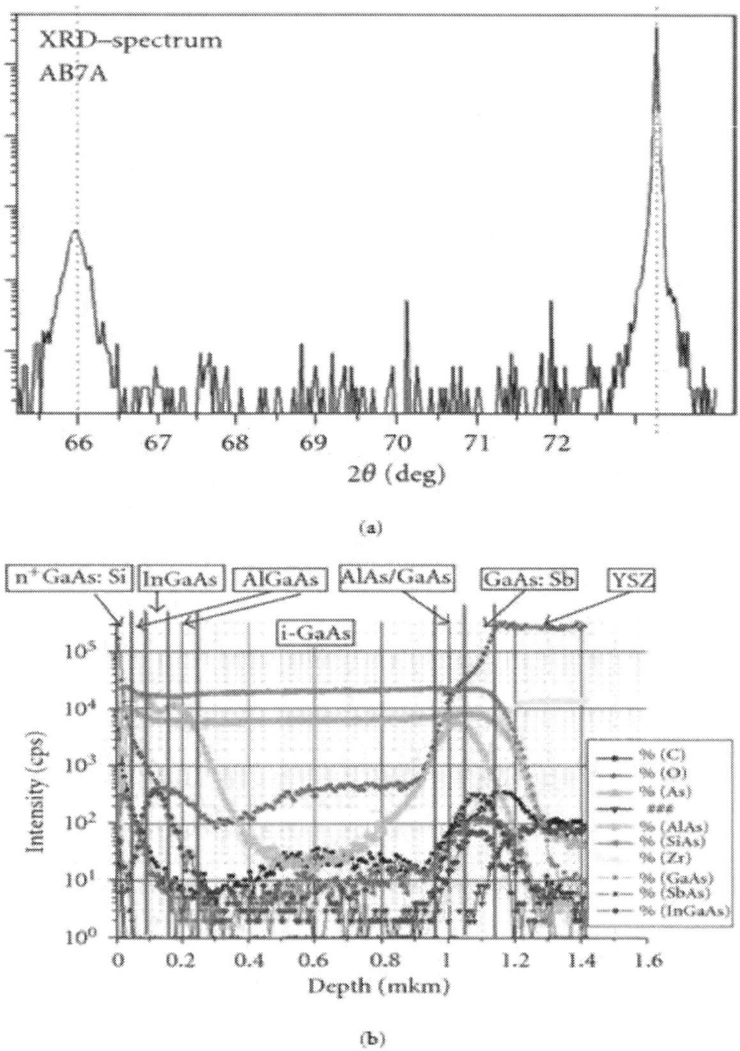

Figure 5: XRD Depth(mkm) scan (a) and layer-by-layer secondary ion mass-spectrometry (b) of AlGaAs/InGaAs/GaAs (001)/fianite (001) multilayer heterostructure.

Detailed data on elemental and molecular compositions of the heterostructures were obtained by means of layer-by-layer SIMS ("TOF SIMS-5" spectrometer). The sputtering was carried out by Cs^+, 2keV, raster µm, negative ion detection mode, the probe beam Bi^+, 25keV, and depth resolution DZ > 7nm. The analysis of the AlGaAs/InGaAs/GaAs heterostructures obtained on fianite (Figure 5(b)) has shown that its inner topology was in conformity with the assigned scheme (Table 2) of the PHEMT structure.

$A^{III}N$ Films on Fianite Substrates and Buffer Layers

Principal difficulty of growth of perfect heteroepitaxial GaN films is an absence of suitable substrates having good matching with the heteroepitaxial film. Currently, for the growth of GaN films, Al_2O_3, ZnO, MgO, SiC, Si, and GaAs substrates are in use. Usually, a material with wurtzite structure is grown on a hexagonal substrate, whereas sphalerite is grown on a cubic one. Fianite as a substrate material for cubic InGaN epitaxy has a number of advantages, such as favorite crystallochemical parameters and high chemical stability. Besides fianite, Si and GaAs substrates with fianite buffer layer were developed in scope of the work. Synthesis of the layer was carried out by a laser deposition technique. The growth of fianite films on silicon substrates was conducted with the purpose to evaluate prospects of the use of less expensive large silicon substrates with fianite sublayer instead of monolithic fianite because maximum dimensions of the silicon-on-fianite structures are limited by size and quality of fianite crystals and the corresponding substrates (currently ~50mm).

Another purpose of the study was determination of suitability of fianite not only as a substrate material but also as a gate dielectric. Production of such substrates will allow integrating GaN-based optoelectronics with a well-developed silicon and gallium arsenide electronics and optoelectronics.

GaN Films on Fianite Substrates Growth of the films on (111) and (100) oriented fianite substrates was carried out using nucleus layers. 3 types of the nucleus layers were used:(1)low-temperature GaN nucleus layer with annealing in hydrogen-ammonia atmosphere;(2)low-temperature AlN nucleus layer with annealing in hydrogen-ammonia atmosphere;(3)high-temperature AlN nucleus layer. At the use of all of the types of the nucleus layers, fianite substrates were annealed in pure hydrogen at ~1070°C before deposition of the films. Hydrogen is a conventional carrier gas in MOGPE of III-V materials because it can be rather readily purified. Similarly, in MOGPE of nitrides of III group hydrogen for the first time was used as a carrier gas. However, later it was demonstrated that in contrast to classic III-V semiconductors, GaN and InN are unstable under hydrogen atmosphere and undergo destruction (etching) at the temperatures used for growth of these crystals. This is an evidence for hydrogen as a carrier gas at the epitaxy of nitrides of III group elements actively participates in the process occurring on the surface of the growing film, in contrast to GaAs. Therefore, in most cases for growth of nitrides of III group by MOGPE, ammonia is used as a nitrogen source and supplied into reactor in large quantities. For a long time, ammonium was of opinion that it inhibits the destruction of a growing film and makes the effect of hydrogen negligible. However, it appears that it is far from the case and hydrogen significantly influences the process of the nitrides growth. The studies have shown that at annealing of LT-GaN nucleus layer, the latter undergo etching in H_2-NH_3 flow hindering growth of a high-quality GaN films. Application of the low-temperature AlN nucleus layer with annealing in hydrogen-ammonia atmosphere as well as the high-temperature AlN nucleus layer on (111) and (100) oriented fianite substrates resulted in formation of hexagonal GaN films comprising a textured polycrystal of hexagonal modification. Scattering angles of the texture for the GaN films grown on the (111) and (100) oriented substrates were 10° and 15°, respectively.

It has been shown that high-temperature annealing of LT-GaN buffer layer at 1000–1100°C promotes improvement of structural perfection GaN heteroepitaxial films. The GaN layers on fianite substrates exhibited an intense photoluminescence with maximum at 365nm.

The conditions of growth of single-crystal GaN films on (111) and (100) fianite substrates by MOCVD without buffer layer at 850°C substrate temperature have been determined. The spectra of $\theta/2\theta$ scanning were obtained using Ge (400) monochromator (Figure 6).

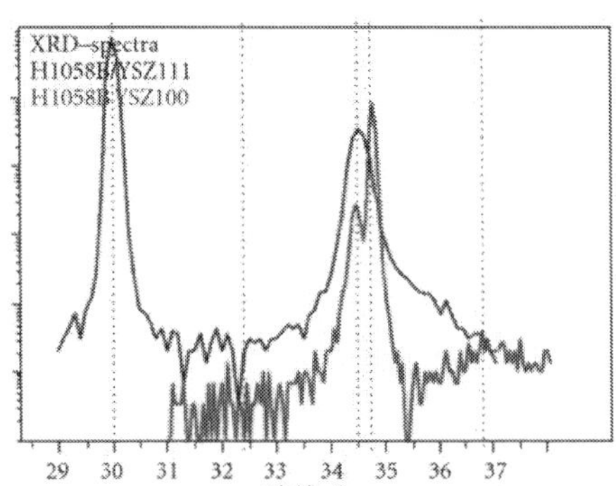

Figure 6. XRD $\theta/2\theta$ scan of GaN films on (111) and (100) fianite substrates.

Two peaks of the substrate were observed at 30° YSZ (111) and 34.8° YSZ (200). The film provides a single GaN (0002) peak at 34.5°. Since GaN (0002) peak is close to YSZ (200) one, a narrow slit in front of the detector was inserted with the purpose to increase the resolution.GaN_{HEX}(0001) was detected on both substrates at FWHM of the XRC<1 that corresponds to the epitaxial growth. Traces of the polycrystalline phase at 32.4° (expected 0.1–1.0 intensity units) were not detected.

AlN Films on Fianite SubstratesThe AlN films on fianite substrates were grown using an MOGPE technique. The $Al_xGa_{1-x}N$ direct gap semiconductors are very useful in the development of UV photodetectors. By altering Al content in GaN-based solid solutions, it is possible to obtain the material with a forbidden band ranging 3.43–6.2eVthus covering 200–365nm spectral band. This spectral band is of practical importance in UV astronomy, ozone layer monitoring, combustion, and water sensors. These films are both of original interest as well as are useful as nucleating and buffer layers in GaN epitaxy. Growth of the films was started from thin 20–50nm nucleating layer. Two growth modes were used: at 650°C with subsequent annealing in ammonia-hydrogen media at 1100°C during 30min followed by growing up of the basic layer and high-temperature growth of AlN at the same temperature. Before the deposition of AlN films, fianite substrates were annealed in pure hydrogen at ~1070°C. Mirror-flat homogeneous AlN films with the roughness not exceeding 0.6nm (Figure 7) were deposited on (100) and (111) fianite substrates. Layer-by-layer analysis of AlN nucleating layer on the fianite substrates was carried out by SIMS using TOF SIMS-5 device (sputtering by Cs+, 2 keV, 250 × 250 raster, negative recording mode, Bi+ probe beam 25keV).

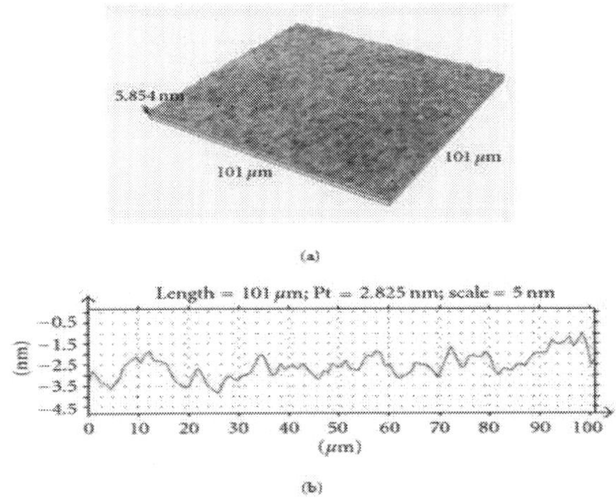

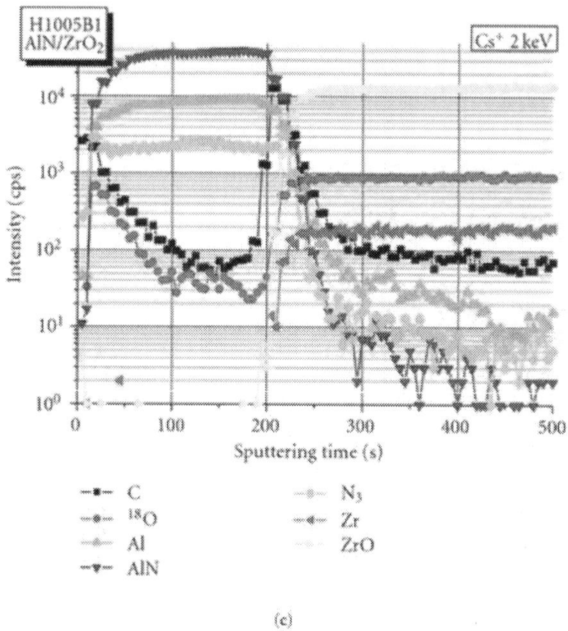

Figure 7. Image of surface (a) and the surface relief (b) (Interference Microscope "Talysurf"); results obtained by layer-by-layer SIMS analysis (c) of the low temperature AlN layer on fianite substrate (TOF SIMS-5).

The study has shown that the films had uniform distribution of its constituents, the concentration profile of Zr atoms at the heterointerface being very sharp (Figure 7(c)). The use of AlN nucleating layers on the fianite buffering layers allows deposition of continuous and homogeneous GaN films of hexagonal modification.

Deposition of InN Single-Crystal Films on YSZ Substrates by Means of LP-MOCVD Technique Using Pulse-Capillary EpitaxyInN heteroepitaxial films attracts a significant interest because the films provide an opportunity of the development of high-frequency electronic devices. Since an industrial technology of InN substrates production is so far absent, heteroepitaxial growth of InN on foreign substrates is required. Al_2O_3 and Si substrates, being well developed and having considerable dimensions along with excellent crystal perfection, seem to be of the most interest. Moreover, Si is not expensive and allows a combination of the InN-based devices with well-developed silicon electronics. However, mismatching between Al_2O_3, Si, and InN lattice constants is rather large, 25% and 8%, respectively. That was the reason why fianite (YSZ) has been recently taken into account as an alternative substrate. For the first time fianite substrates have been successfully tested for deposition of InN heteroepitaxial films in [4]. The films of cubic symmetry were grown on (001) fianite substrates using plasma-assisted molecular beam epitaxy (PA-MBE) at 400–490°C.

Due to better crystal-chemical matching between InN and fianite, the films grown on fianite substrates exhibited crystal perfection better than those grown on sapphire and GaAs substrates [4]. In [5], InN epitaxial films have been for the first time deposited on fianite substrates by pulse-capillary LP-MOCVD. The technique was shown to allow a significant improvement of the films quality, as well as synthesizing single-crystal InN films of hexagonal modification. The films demonstrated rather perfect morphology along with high photoluminescent and electrophysical characteristics. The structures on fianite substrates were grown under low pressure in vertical reactor supplied with rotating molybdenum stand (designed in IPM RAS). The substrates were inductively heated, trimethyl indium (TMIn) and ammonia were the element sources, and nitrogen was the carrier gas. The substrates of (111) and (100) orientations were used. Comparative studies of the InN films were carried out using optical and electron microscopy, SIMS, XRD, photoluminescence, and probe techniques. The experiments on deposition of InN films on fianite by MOCVD technique have shown that it was difficult to grow a uniform and smooth InN film using general growth process, as well as single-step capillary epitaxy, because of three-dimensional nucleation. Satisfactory results have been achieved only by using pulse-capillary epitaxy, in which the process of intermittent supply of In (in form TMIn) at constant NH_3 flow was repeatable [7]. The pulse-capillary LP-MOCVD deposition was started by puffing TMIn in nitrogen without ammonia inflow during 20s followed by 40s interruption of TMIn supply and puffing nitrogen-ammonia mixture. TMIn flow was equal to 0.25sccm/min. The procedure was repeated ten times and then the growth of InN film was conducted at 550°C in N_2-TMIn-NH_3 gas mixture. The V/III reactant ratio was in the range of 10^5–10^6. Under these conditions growth of InN films of 0.3μ thickness was observed. The application of "pulse-capillary epitaxy" technique and fianite substrates resulted in the formation of mirror-flat films of single-crystal structure. The comparative studies on growth (nucleation and formation of the primary continuous layer) of InN films on fianite substrates by means of LP-MOCVD and conventional MOGPE techniques have demonstrated significant differences in size, habitus, and concentration of the epitaxial nuclei, as well as in morphology of the epitaxial layers. Raster electron microscopy (Supra 50VP) images of the surface of InN layers deposited by means of conventional LP-MOCVD (a) and pulse-capillary LP-MOCVD techniques (b) are shown in Figure 8. Roughness of the surface of InN layer of 0.3μ thickness grown by the conventional technique is apparently seen. The roughness occurred because of the three-dimensional nucleation.

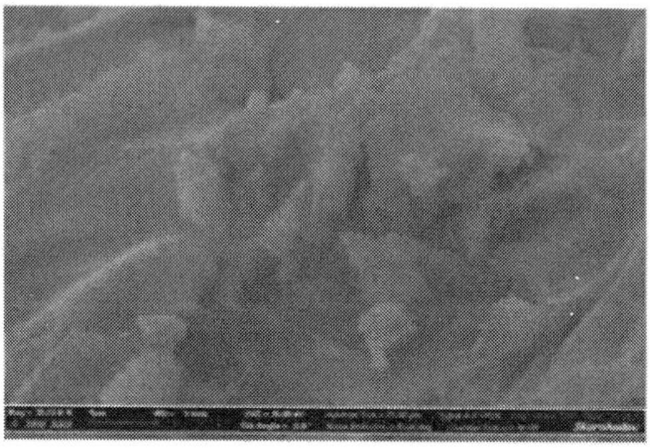

(a)

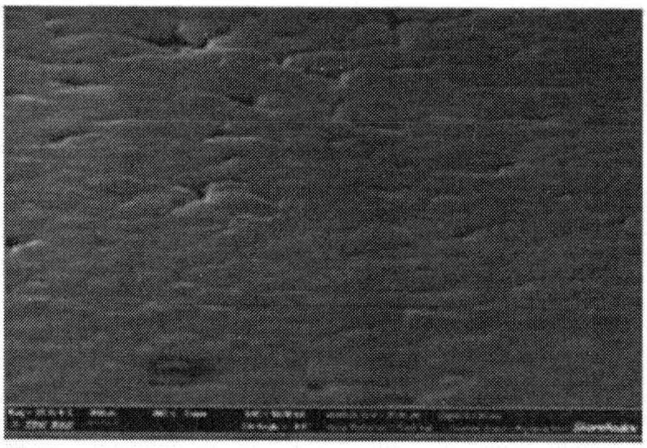

(b)

Figure 8. Morphology of the surface of InN films deposited by means of conventional (a) and pulse-capillary LP-MOCVD techniques (b) (scanning electron microscope Supra 50VP).

At the same time, pulse-capillary epitaxy resulted in more smooth and uniform surface. $\theta/2\theta$ scanning spectrum of InN film deposited on (111) fianite substrate by pulse-capillary LP-MOCVD technique and rocking curves of InN films are presented in Figure 9. The 2θ peaks in the spectrum are attributed to 30.0—YSZ (111) substrate, 31.3—InN (0002) layer, and 33.0—In (100).

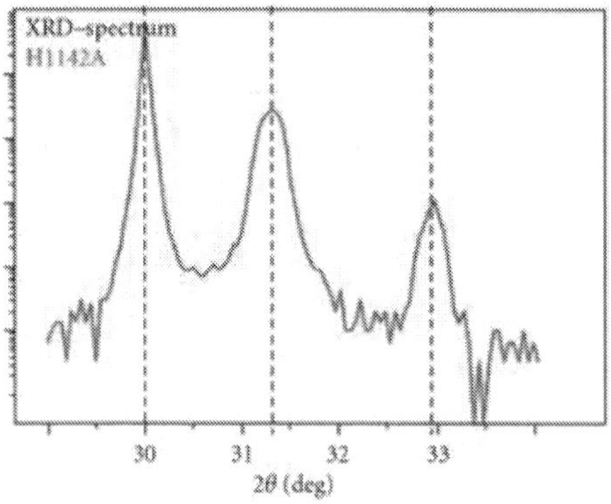

Figure 9. XRD $\theta/2\theta$ scan of the InN film.

The rocking curve of InN (0002) film deposited on YSZ substrate by means of pulse-capillary epitaxy is shown in Figure 10. Rocking curves of InN films deposited on YSZ and sapphire substrates by means of conventional techniques are shown for comparison. The rocking curve data of InN (0002) peak provide full width at half maximum 1,8° for the films grown on sapphire (curve 3 on Figure 10). The values obtained for fianite films grown using conventional and LP-MOCVD using pulse-capillary epitaxy are 1.0° and 0.19° (curves 2 and 1 on Figure10), respectively. It is noticeable that the latter value is considerably lower. Therefore, the growth on fianite results in more structurally perfect InN films due to better crystallographic matching, other conditions being equal. Herewith, application of the pulse-capillary epitaxy allows achieving of record values that is the evidence of excellent perfection of InN films obtained in this study.

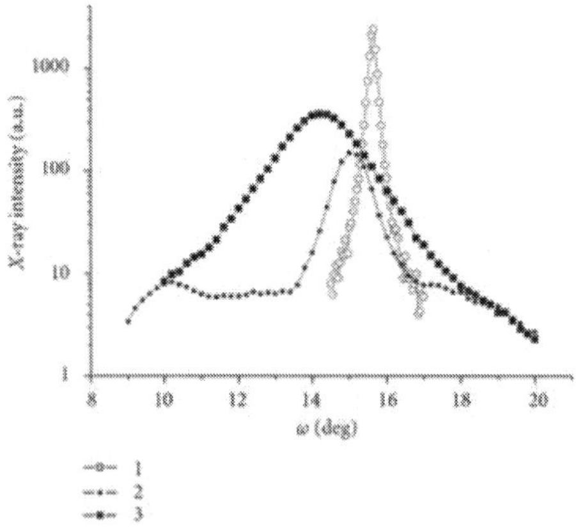

Figure 10. Rocking curves of (0002) InN.

The nondoped epitaxial InN films grown on fianite substrates were semiconductors of n-type with 2.5 * $10^{19}cm^3$electron concentration. It is also noticeable that Hall mobility value approaches $320cm^2/V \times s$. The films showed intense photoluminescence with peak maximum at 0.87eV.

Electrically Active Defects in GaN Films on GaAs Substrates with Fianite Buffer Layers

Comparative studies of density and electric activity of structural defects in GaN epitaxial films grown on GaAs substrates with various buffer layers were carried out by an induced bias technique (IBT). The technique has been developed rather recently [27, 28]. It is a contact-free similarity of the induced current technique (EBIC-mode). IBT is a nondestructive contact-free method for diagnostics of semiconducting materials and microelectronic devices. IBT is based on detecting voltage (or charge) generated by an electron probe of scanning electron microscope (SEM). Draft scheme of the method is shown in Figure 11(a).

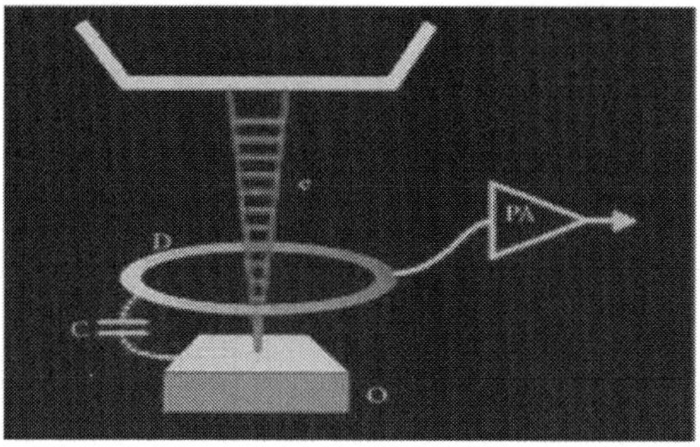

(a)

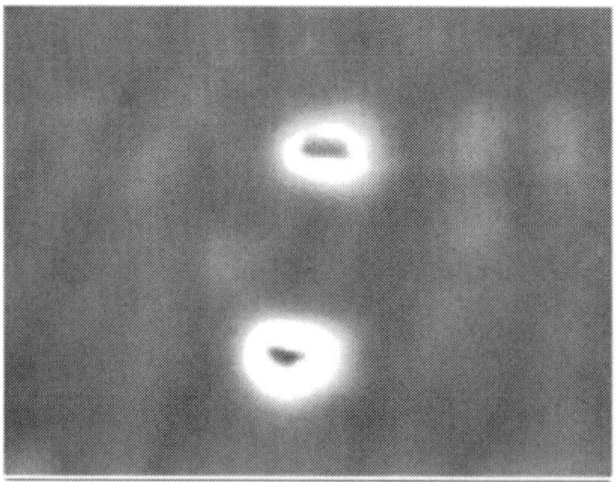

20kV x10000 1 μ

(b)

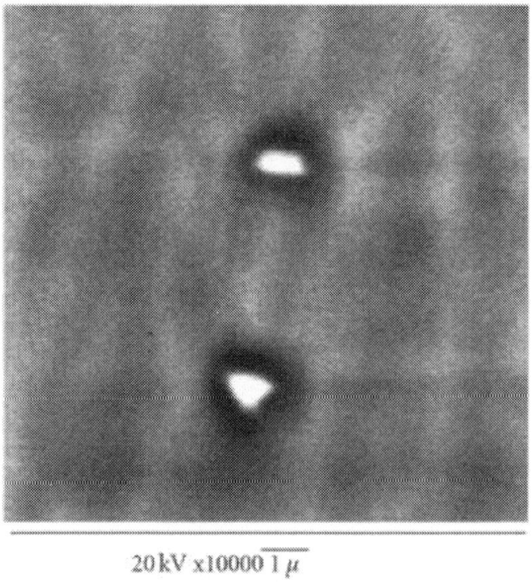

(c)

Figure 11. Outline of the induced potential technique (a) and scanning electron microscope images of electrically active polygonal defects in GaAs films: secondary-electron emission mode (b); b-induced potential mode (c).

The electron probe (e) scans the surface of a crystal under the study (O). Metal ring (D), in which the surface charge generated by electrons through capacitive coupling is induced, is a detector of a signal. The signal from the ring electrode is monitored in the SEM display (or by other measurement equipment) through charge-sensitive amplifier (PA) (Figure 11(a)). The technique allows qualitative monitoring of semiconductor plates, structures, and devices identifying electric active inhomogeneities such as dislocations, stacking faults, microfractures, extent of doping by various dopants, all n-junctions, and Schottky barriers (see e.g., Figures 11(b) and 11(c)). Quantitative measurements of local fundamental characteristics of semiconductors are also possible (diffusion distance, nonequilibrium carrier lifetime, its surface recombination rate, diffusion barrier height).

The studies have shown that the use of GaAs substrates with porous GaAs layer resulted in a decrease of the electric activity of structural defects in the GaN films and in an increase of its electrical uniformity as compared to GaN films grown on monolithic GaAs substrates. The use of GaAs substrates with double buffer layer (fianite on porous GaAs) allows additionally decreasing concentration of the electrically active defects in GaN films to more than an order of magnitude (Figure 12).

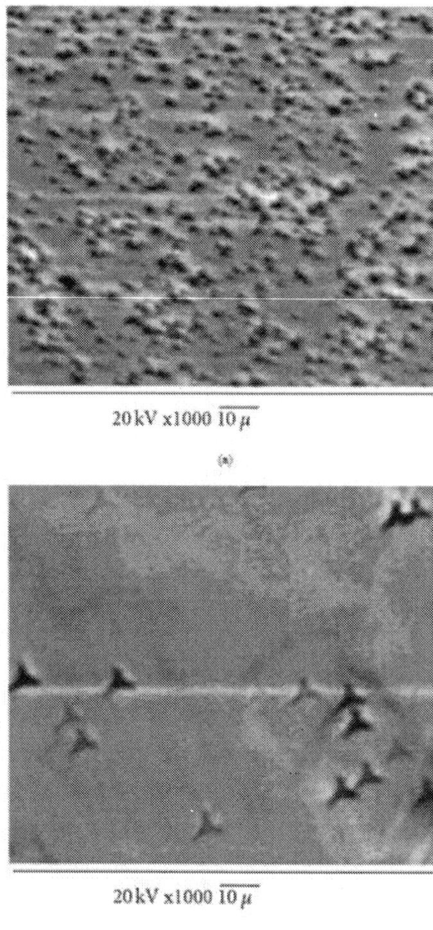

Figure 12. Electrically active defects in GaN film on GaAs substrate with buffer layers: (a) single buffer (fianite); (b) double buffer (fianite on porous GaAs).

GaN Films on Si and GaAs Substrates with Fianite Buffer Layers

Silicon and gallium arsenide are promising substrates for GaN and the other $A^{III}N$ epitaxy due to their high quality, large dimensions, and a low net cost, as well as possibility to integrate GaN-based devices with well-developed silicon and gallium arsenide electronics and optoelectronics. However, there are three considerable problems occurring at GaN epitaxy: first, a significant parameter mismatch of GaN layer and Si or GaAs substrates; second, the difference of thermal expansion coefficients of the layer and the substrates and third, insufficient chemical and thermal stability of the substrates at the epitaxial temperature. Application of various buffer layers, in particular, fianite-based, can be an efficient method for solution of the above problems.

GaN epitaxial films were grown by MOCVD technique using capillary epitaxy on Si and GaAs substrates with various buffer layers. Trimethylgallium (TMG), arsine (AsH_3), and ammonia (NH_3) were used as Ga, As, and N sources, respectively. Single (fianite, layer of porous Si, or GaAs material) and double (fianite on porous Si and GaAs) were tested. The first "prominent" porous buffer layer was suggested to allow decreasing thermoelastic strains in the second heteroepitaxial buffer thus improving its morphology and structure. The upper buffer layer, being chemically stable in the growth medium, provides fine matching with functional heteroepitaxial film.

The epitaxial structures were studied using a set of techniques: photoluminescence (PL), scanning electron microscopy in electron-beam induced current (EBIC-mode), and secondary-ion mass spectroscopy (SIMS).

It was established that the use of fianite buffer layer on Si substrate prevents formation of amorphous silicon nitride. The GaN films grown on Si substrates with fianite buffer layer were of hexagonal modification and had mosaic single-crystal structure. It was demonstrated that the use of porous Si in the complex fianite/Si buffer allows improving the adhesion of GaN film and its uniformity by phase composition and thickness.

Layer-by-layer SIMS analysis of GaN films grown on Si and GaAs substrates with fianite buffer layers has shown that fianite layer serves as a barrier for diffusion of Si and As into GaN film from Si and GaAs substrates, respectively (Figure 13). Good insulating properties of ZrO_2 in the double buffer provide an opportunity to use "Semiconductor on dielectric" technology, which is promising to improve the integration level.

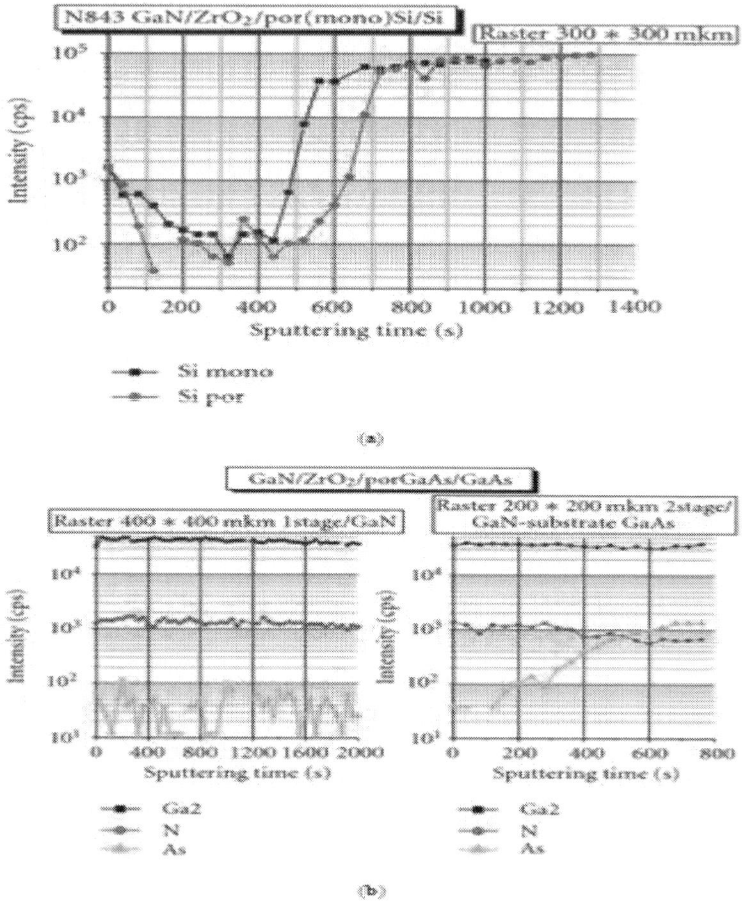

Figure 13. Results from a layer-by-lyer SIMS analysis of GaN/fianite/por(mono)Si/Si (a) and GaN/fianite/porGaAs/GaAs (b) structures.

Comparative studies of PL spectra (recorded at 300°K) of GaN films grown on a monolithic GaAs substrate and GaAs substrates with various kinds of buffer layers have been carried out (Figure 14): 1 single buffer "porous GaAs;" 2 double buffer "fianite on porous GaAs."

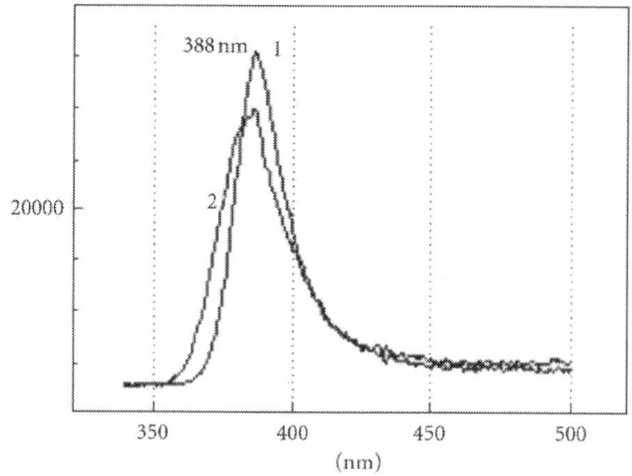

Figure 14. Photoluminescence spectra of GaN films (300°K) on GaAs substrate with buffer layers: porous GaAs (1) and double buffer-fianite on porous GaAs (2).

The position of PL peaks in the spectra corresponds to characteristic peak of cubic GaN. Consequently, the use of single buffer layer of porous GaAs as well as double buffer layer (fianite on porous GaAs) allows growing GaN films of cubic modification. The growth of GaN film on monolithic GaAs substrate in contrast resulted to the formation of hexagonal modification.

FUNCTIONAL FIANITE FILMS

Functional Fianite Films on Si, Ge, and GaAs Substrates

Thin films of fianite and related solid solutions such as $Zr(Ce)O_2$ can be used as insulating layers (alternative to SiO_2, SiC, and Si_3N_4) in the development of Si-, Ge- and GaAs-based "semiconductor-dielectric" multilayer structures. Fianite is also a good gate dielectric for Si- as well as for $A^{III}B^{V}$ -based devices (including GaN-based) due to its high dielectric constant value (25–29.7). Thin fianite films are a barrier for diffusion of impurities and provide a significant (up to 1000-fold and even more) decrease of the loss current in highly integrated devices [11,12]. Due to high chemical inertness fianite films can also be used as protective coatings.

Techniques for Deposition of Fianite Films on Si and GaAs Substrates

In recent years, a considerable attention was drawn to fianite films on silicon due to its electric and optic device applications, such as insulating layers in SOI (silicon-on-insulator) devices [29], gate dielectric in Si- [28, 29], SiGe- [30], and $A^{III}B^{V}$ -based [31] device structures, buffer layers for production of optic coatings for films of various semiconductors [30–33], superconductors [34–36], ferroelectrics.

Various techniques can be used for the production of fianite films on silicon and other semiconductors, including magnetron [36–40], laser and electron-beam [41–43] sputtering, and molecular-beam epitaxy (MBE), as well as gas-phase chemical deposition [44].

The choice of a specific technique is determined by further designation of a fianite film and possibility to produce the film of maximum structural perfection, as well as technologic potentialities of the technique. So, MBE technique is more suitable for the deposition of the thinnest fianite film for the use as a gate dielectric. Magnetron and laser sputtering are more favorable for fianite layers used as buffer layers with subsequent growing semiconductor films, including $A^{III}B^{V}$ compounds. In [36], fianite films were deposited on Si and GaAs substrates using magnetron, laser, and electron-beam sputtering techniques. The films obtained by magnetron sputtering were of the best structural perfection [36].

Growing of Fianite Films on Si and GaAs Substrates

The growth of fianite films on silicon and gallium arsenide substrates was carried out with the purpose to evaluate the prospects of using less expensive and more large Si and GaAs substrates with fianite sublayer instead of monolithic fianite substrates because, currently, maximum size of the latter is ~50mm. Another purpose was the determination of an opportunity to use fianite not only as a substrate but also as insulating layer material alternative to SiO_2, SiC, and Si_3N_4 protecting and insulating layers, as well as a gate dielectric for multi-layer "semiconductor-dielectric" structures. Producing such substrates will allow integrating GaN-based optoelectronics with a well-developed silicon and gallium arsenide electronics and optoelectronics. Magnetron and laser sputtering were used for deposition of fianite films on silicon and gallium arsenide

Fianite Films on Porous Si and GaAs

With the purpose to improve quality of fianite films and its adhesion to Si and GaAs substrates, opportunities of the use of porous layers of the material were studied.

The following results were obtained:

i. Appropriate regimes of deposition of porous GaAs layer on GaAs (111) substrates of n- and p-conductivity types were developed;
ii. Appropriate regimes of deposition of the uniform mirror-flat fianite layer on GaAs (111) substrates of 18 × 18mm size were established;
iii. It has been demonstrated that the use of the porous layers allowed an improvement of adhesion of fianite with GaAs layers;
iv. The samples of fianite/GaAs and fianite/Si epitaxial substrates have been obtained for subsequent growth of $A_{III}N$films.

High mechanical and chemical stability of fianite and absence of pores open the prospects of its application as protective and stabilizing coating for various substrates.

Magnetron Sputtering Technique

Magnetron systems are related to diode-type sputtering systems. The sputtering occurred due to bombardment of a target surface by gas ions (usually Ar) forming in plasma of anomalous glow discharge. Material ions knocked out the target subjected to the bombardment are captured by the magnetic field and maintained complex cycloidal movement by closed trajectory in vicinity of the target surface. High sputtering rate, which is a feature of magnetron systems, is achieved by an increase of the ion current density due to localization of plasma by means of high transverse magnetic field. The increase of sputtering at simultaneous decrease of actuation gas pressure allows a significant decrease of contamination of the films by alien gas impurities. Fianite was grown up on Si and GaAs substrates using unbalanced magnetron system. Fianite crystals were used as a target. Si substrate subjected to the sputtering was heated by IR radiance. Preparation of the substrates included degreasing, removing of the oxide, and passivating of the surface in ammonium-peroxide solution. Optimization of the conditions of the growth of fianite films on Si substrates was carried out by varying of the sputtering rate, temperature of the substrate, and residual gas pressure.

Bombardment of the target leads to dissociation of zirconium and yttrium oxides to ZrO, Zr, YO, Y, and O_2. That is why such parameters as sputtering rate and residual gas pressure considerably influence stoichiometry of the resulting film. Energy of the evaporating particles is rather low (~0.5–10eV); so for the epitaxial growth of fianite film, a high temperature of Si substrate and optimal rate of the condensate supply are necessary.

The fianite films were studied by means of scanning electron microscopy, ellipsometry, and CV-parameters measurement techniques. The film parameters were found as follows:

i. Optic refractive index n_d~ 2.1÷2.2;
ii. Dielectric constant ;
iii. Absence of defects of porosity type (in 30mm diameter sample).

Laser Sputtering Technique

Experimental installation for deposition of fianite films was a sputtering system composed by a vacuum device and excimer laser. The system has been designed and manufactured in IPM RAS.

Operational oxygen pressure was maintained by vacuum system supplied with a mechanical pump and CHA-2 letting system. Evaporation of the target was performed by LPX200 eximer laser radiation working on KrF mixture. Wavelength of the radiation was 248nm, pulse duration 27ns, the pulse energy 350MJ (pulse power 1.3×10^7W), and repetition frequency 50Hz. Optical system providing a focusing of the laser beam on the target surface consisted of quartz prisms and 30cm focal distance lens. The laser beam spot on the target surface was $1 \times 4mm^2$. The energy density on the target surface was ~$10J/cm^2$. The distance between the target and substrate was 60mm. Cylindrical targets of 15–20mm diameter and 10–30mm length were used in the installation. In order to prevent local overheating of the target and to provide a uniform material drift rotation an axial movement of the target was used. Possibility of conducting pre- and postgrowth annealing under oxygen atmosphere at 10Pa–100kPa pressure and at up to 750°C temperature is a peculiarity of the installation.

Ceramic target of $(ZrO_2)_{1-x}(Y_2O_3)$ with x=0.1 composition was used for deposition of fianite films. The deposition was carried out on Si and GaAs substrates heated to 600–800°C temperature under oxygen atmosphere at approximately 10Pa. The growth rate of YSZ films was about 0.02nm per pulse. Contactless heater of substrates (heating by irradiance) was an original peculiarity of the sputtering system. The heater comprises vertically positioned quartz tube (of 30mm inner diameter) supplied with refractory stainless steel heating coil on its outer surface with up to 1kW power of the heater. Monitoring and maintenance of the assigned temperature (with ±5°C precision) were carried out using precise regulating device and Pt-Rh thermocouple positioned under the heating coil. A substrate was fitted in a holder and positioned inside of the quartz tube. Loading of substrates and oxygen supply was maintained through the upper end of the tube.

Technology of growth of dielectric fianite films using the laser sputtering consists of the following stages.(1)A substrate is loaded to the sputtering system and the vacuum chamber is evacuated up to ~1Pa residual pressure.(2)Letting-to-oxygen is maintained up to the pressure required.(3)Rotational movement drive of the target is switched on.(4)A substrate is heated up to deposition temperature.(5)The eximer laser (the pulse energy 350MJ, repetition frequency 50Hz) is switched on and the sputtering is started.(6)Followed by the achievement of the assigned thickness of the film, the laser is switched off.(7)Followed by the end of the film growth, the chamber is filled with oxygen up to the pressure required.(8)The structure is annealed.

The substrate heater is switched off and the substrate is cooled to room temperature.

Initial Stages of Deposition and Structure of Fianite Buffer Layers on Si and GaAs Substrates

The application of fianite as a buffer layer will present a solution route to another very important problem: epitaxy of $A_{III}N$ compounds on Si and GaAs substrates having large dimensions, high quality, and low net cost.

Single-crystalline heteroepitaxial fianite films of 1000A° thickness were grown on silicon substrates of up to 50mm diameter in vacuum chamber at p~ 210^2Pa pressure, sputtering rateV_S~ 60/min, and substrate temperature A° ~ 800°C.

The studies have shown that the film became continuous as from 100 A° thickness.

X-ray structural studies of ZrO_2-Y_2O_3/Si structures have shown that the fianite film was single phased and consisted of two layers with different rocking curve values: 0.20° for the upper layer and 0.96° for the lower one. Epitaxial relation between the film and the substrate was (100)[100] Si//(100)[100]ZrO_2-Y_2O_3. The relation was established using diffraction measurements under the following regimes: $\Theta/2\Theta$ scanning (simultaneous rotation of the detector and sample over goniometer axis) and Ψ -scanning (rotation of the plate in a proper plane at fixed detector position). The former regime $\Theta/2\Theta$ was used to determine orientation of the composition plane, the latter-mutual orientation of unit cells of the film, and the substrate in the composition plane.

Spectra of the -scanning of $(ZrO_2-Y_2O_3)/Si$ structure for asymmetric (422) reflection of the film (b) and the substrate (a) are shown in Figure 15.

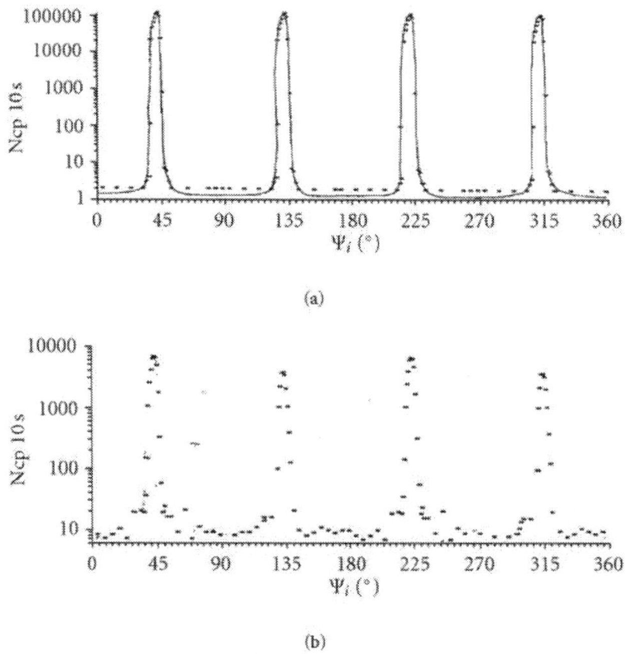

Figure 15. Ψ scanning spectra for (422) reflection of YSZ substrate (a) and the film (b).

The absence of additional peaks and high peak maximum-to-background ratio (~10^3) are the evidence for ZrO_2-Y_2O_3 layer is a perfect single-crystal film. The fianite buffer layers grown on Si and GaAs were used for compounds epitaxy.

Some Difficulties in Deposition of Fianite Films on Silicon Substrate

Growth of fianite-on-silicon structures of high quality featuring with sharp interfaces is associated with significant difficulties because of a number of principal problems.

First, silicon surface readily undergoes transformation into SiO_2 amorphous layer either due to interaction with oxygen-containing fianite film or oxidative atmosphere usually used at fianite growth. As it has been shown by calculations, fianite should not react with silicon substrate to form SiO_2, which has low dielectric constant value, at a direct contact [45].

However, in practice, it is very difficult to avoid the formation of this film at fianite deposition or subsequent high-thermal treatment [46, 47]. Therefore, a development of special technological tools is necessary. One of the routes to solve the problem has been suggested by the authors [48]. Thin Zr or Y layer was deposited on Si substrate before fianite deposition. The metals absorb oxygen from SiO_2 layer because free energy of both fianite and Y_2O_3 formation is lower than of SiO_2 one [49]. That leads to a decrease of the layer thickness.

Second, oxygen from fianite layer readily diffuses to silicon substrate or reacts with silicon surface. Secondary phases occurring as a result of the reaction disturb silicon crystal lattice and hinder a perfect growth. Under these circumstances, fianite layers on Si substrates are of amorphous or polycrystalline structure. At the development of gate dielectric technology, these issues are of peculiar importance because thickness of the last layer is about some nanometers.

Therefore, the above data show that the problem of deposition of fianite films on Si substrates is of great interest. The problem of improvement of quality of the layers seems to be very urgent because of a number of principal difficulties occurring due to peculiarities of physical-chemical properties of the materials resulting in reactions at the growth and subsequent thermal treatment stages. The synthesis of perfect fianite films on Si requires a development of special methods to decrease the influence of amorphous SiO_2 layer at the substrate-layer interface.

Characterization of Fianite Films

The films were studied by means of scanning electron microscopy, ellipsometry, and CV-parameters measurement techniques.

Capacity-Voltage Characteristic Measurements of the "Fianite-on-Silicon" Structures

The capacity-voltage (CV) characteristics of the structures supplied with fianite films deposited on p-Si and n-Si substrates were measured.

Capacity measurements provide evaluation of dielectric properties of films under the study: dielectric constant and dielectric loss tgδ. The application of a multifrequency device allows the determination of frequency dependencies of dielectric constant and high-frequency loss in dielectric films. Since the dielectric film is deposited on a semiconductor, a MIS structure (metal-insulator-semiconductor) is formed, so the CV-measurement provides additional information concerning the semiconductor and the dielectric-semiconductor interface, namely, type of the semiconductor conductivity (n or p) and concentration of the dopant, flat band barrier voltage V_{fb}, density of boundary states, and a charge induced in the dielectric.

The device used for CV-measurements allowed determining the capacity and high-frequency conductivity of the structures, as well as its dependency on the applied voltage. The measurements were carried out at 500KHz and 1MHz frequencies. Direct potential bias range was ±40V. Thermally sputtered Al of 1mm surface diameter was used as the contacts. The results obtained are shown in Figure 16.

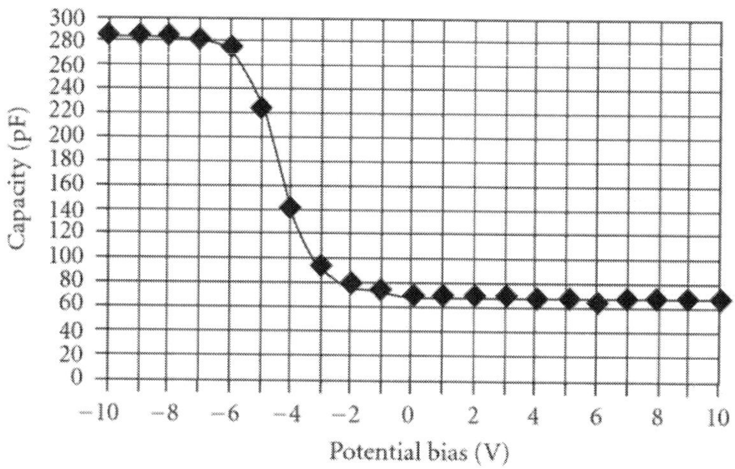

Figure 16.CV-characteristics of fianite-on--Si sample.

"Al-fianite-Si" MIS structure parameters: flat band barrier voltage –4V for 180nm film and 1.5V for 20nm film; density of the boundary state charge ~+10^{12}cm^{-2}.

Investigation of ZrO_2 Films on Si and Ge Substrates by Means of Scanning Electron Microscopy

The ZrO_2 films were studied using scanning electron microscopy. All of the films studied were porous-free. Since the square of the samples studied was 5-6 cm^2, it is possible to consider the porosity value at least not exceeding 0.15–0.2cm^{-2}. For comparison, it is worth to mention that porosity of SiO_2 films is 4–8cm^{-2}. Therefore, it is possible to consider ZrO_2 films as a protective layer for Ge devices actually superior SiO_2 films because its porosity decreased in 1.5–2 orders of magnitude.

The study of morphology of the films deposited by magnetron sputtering technique at high magnification has shown its satisfactory homogeneity. Some regions of the surface featured by a relief composed by quasi-spherical hills of 500–600nm in diameter and exhibiting lateral periodicity. Analytical study of the films has shown an absence of inclusions of impurities.

An attempt to study the mechanism of formation of the films with the purpose to optimize the conditions of magnetron sputtering was done using electron microscopy (JSM JEOL 5910 LV). The particles were identified by means of electron probe. The film was removed by polishing using diamond paste with 2.5–4μm particle size. This abrasive size was chosen to minimize decreasing particle size of the film constituents at the polishing. The obtained material was flushed by ethanol (9–12 purity grade "for microelectronics") and the suspension was put in plastic syringes (1mL). In order to disintegrate aggregates, ultrasonic (US) treatment was carried out. The US dispersion was conducted using "Sapphire 3M-1.3" US device with 35GHz operational frequency. The syringes were inserted to the device chamber filled with water. The chamber was thermostated at 27°C. Followed by 3min of the US treatment, the suspension was aspirated onto conductive (graphitized) ribbon for subsequent microscopy study. The study has shown that the largest constituents of the zirconia film were quasi-spherical particles of 50–100nm size that explained X-ray amorphous nature of the film. It is possible to suggest that formation of larger elements of the relief occurred by enlargement of such particles. The reasons of local enlargement (formation of spherical hills) can be gradients of temperature and mass transfer, as well as occurrence of impurities. The observations allowed refining the conditions of the sputtering of ZrO_2 and fianite films in order to minimize surface roughness.

Fianite as a Gate Dielectric

Recently, a sharp surge of interest in the use of fianite as a gate dielectric in CMOS technology has been observed. It is associated with an increase of leakage current at the use of conventional SiO_2 by increasing the integration level. That requires a change of SiO_2 over dielectrics with higher dielectric constant (high-k materials) [30, 32, 33,45]. The resent studies have limited possible alternatives to fianite, HfO_2, ZrO_2, and its silicates. For example, ZrO_2 has high dielectric constant value, good dielectric properties (5.8eV energy gap width), and rather good crystallochemical matching with Si [50] (see Table 2). Intel Corp., one of the leaders of the world electronics, has demonstrated that the change of SiO_2 over HfO_2 as a gate dielectric in 45nm technological process allows decreasing leakage currents, (which became a serious problem for transistors) by more than two orders of magnitude [51].

Comparison of fianite films [33] and SiO_2 with electrical equivalent oxide thickness of about 1.46nm, has shown that the leakage current for fianite was four orders of magnitude lower than that of conventional SiO_2 gate oxides.

The hysteresis and interface state density in this film was measured to be less than 10mV and $2.0 \times 10^{11} eV^{-1}cm^{-2}$. Thus, such crystalline oxide on semiconductor could be used for future generation of semiconductor-based devices.

It is worth to note that quality of the synthesized fianite, as well of the interfaces, is very important for integration of such a dielectric to the CMOS technology currently in use.

Synthesizing of fianite-on-silicon structures of high quality featuring with sharp interfaces is associated with significant difficulties described above. One of the routes to solve this problem is in application of low-temperature growth and annealing regimes, as those, which were used in the series of experiments described below, type of a substrate, and the annealing media were also varied. Conditions of the synthesis of the fianite/Si structures are given in Table 3. XRD technique has shown that fianite layers obtained by laser deposition at room temperature were of amorphous structure.

Table 3. Parameters of growth and annealing of the fianite-on-Si films.

Sample	*T* of growth, C	Annealing, 600 C, 10 min	Film thickness, nm	Substrate
z1	Room	Without annealing	~20	Si (B)
z2	Room	Vacuum	~20	Si (B)
z3	Room	Oxygen	~20	Si (B)
z4	Room	Oxygen	~20	Si (Sb)
z5	600	Oxygen	~20	Si (B)

Subsequent postgrowth recrystallization annealing resulted in arising of a polycrystalline phase in the film. At the same time, the layers sustained mirror flat and uniform. Profile of the surface of z4 sample (Table 4) obtained using Talysurf interference microscope is shown in Figure 17(a). Roughness of this ZrO_2 surface was estimated as Sq = 0.852nm that does not practically differ from roughness of the Si substrate used for the fianite growth (Sq = 0.7877nm).

Table 4. Optimal modes of fianite and ZrO_2 sputtering

Target material	Fianite	ZrO_2
Installation	Z-400	Z-550
Target diameter	Ø 70 mm	Ø 203 mm
Argon pressure	$5 * 10^{-3}$ mBar	$5 * 10^{-3}$ mBar
Power	~500 Wt	~400 Wt
Film growth rate	100 Å/min	~50 Å/min

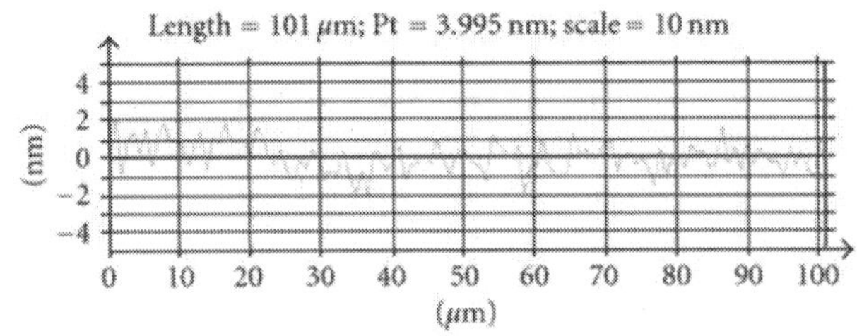

(a)

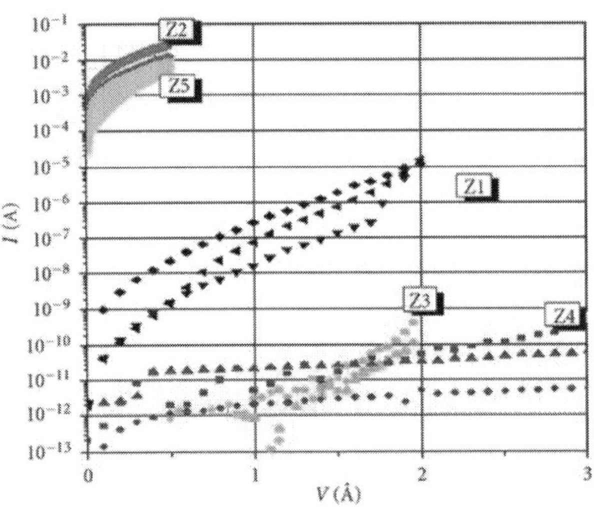

(b)

Figure 17. Surface roughness of fianite film on Si substrate, sample z4 (a); and leakage current of Al/fianite/Si structure (b) samples z1–z5 were prepared under different conditions.

Preliminary studies of gate properties of thin (10–15nm) fianite films obtained by laser deposition on Si substrates have been carried out. The studies conducted on the test structures with deposited Al contacts have shown that thin fianite films featured with low values of loss currents, minimum values being 10^{-12}A/cm^2 at 1V voltage (Figure 17(b), samples z3 and z4).

Fianite and ZrO_2 as Protective and Stabilizing Layers on Ge and Si Substrates and Multilayer Structures

Deposition Modes

Fianite and zirconium dioxide films were deposited by magnetron sputtering technique, 2 types of vacuum evaporation Leybold Heraeus units (Z-400 and Z-550) with different target dimensions were used: 70mm in diameter for fianite and 203mm—for ZrO_2 (Table 4).

High-frequency and direct voltage modes of sputtering were tested. By using the latter mode it was impossible to provide sufficient film growth rate, so HF sputtering mode (13.56MHz) was chosen. The optimal modes of fianite and ZrO_2 sputtering are also shown in Table 4. In case of low magnetron power, plasma is unstable ("blinking plasma"); in case of larger values of discharge power, the growth rate increases, but irregularity of the surface patterns and growing film coarse-graining were observed on a number of samples. Fianite sputtering requires higher power than in case of ZrO_2, at the same time the growth rate was twice as much than in case of ZrO_2.

The developed technique of magnetron sputtering made it possible to vary the fianite film thickness between 600 and 2000 A°. The optimization of the sputtering modes allowed producing of Ge and Si plates with fianite films of satisfactory quality. Ge plates with fianite films were used to try out further operations of the device structures production: photolithography and etching.

Protective and Stabilizing Properties of Fianite Films on Ge

Inorganic dielectric coatings are usually used for passivating and protection of p-n transition surface, as shielding and thermal compensation layers at ion implanting and for interference antireflecting protection. Passivation of the surface is the most important issue for manufacturing of germanium photodiodes because natural GeO and GeO_2 oxides are unstable and, so, cannot be considered as the only passivating coatings. It is one feature distinguishing Ge and Si devices (the latter have stable and rather efficient coating formed by its own SiO_2 oxide). This oxide film deposited from a gas phase is of the most frequent use for photodiodes, with p^+N^- structure. It has positive charge and by attracting electrons to the surface prevents growth of p-channels thus decreasing probability of generation in the layer. It is worth to note that for improved reliability and stability of characteristics of photodiodes, it is necessary to maintain surface state density at $10^{11}cm^{-2}eV^{-1}$ level. However, this passivating technique is far from ideal because of the high porosity of SiO_2 films that decreases humidity resistance and reliability of the device.

In order to improve dielectric properties of the protective coating, fianite films deposited by magnetron sputtering were used. The opportunity of its application for maintaining high-quality practically porous-free protective coating has been confirmed earlier by the experiments.

It has been demonstrated that the use of fianite protective layer in Ge-structures instead of SiO_2 eliminated pulse noise and thus considerably improved photoelectric and performance characteristics of these devices. It has been established that the improvement was related to a more uniform nature of fianite films, in particular, absence of pores, in comparison with SiO_2 films, which contain defects in form of pores.

Some Properties of the Device Structures Supplied with Zirconium Dioxide Films

Photoelectric characteristics and noise of germanium photodiodes supplied with ZrO_2 and SiO_2 films described above have been investigated. Monochromatic sensitivity of these photodiodes is typical for germanium devices and equals to 0.5-0.6A/W (at 1.06 and 1.55μm wavelengths). The change SiO_2 over ZrO_2 resulted in somewhat decrease of a dark current (on average for 10%). Main improvement of the photodiodes quality achieved due to the application of ZrO_2 films was revealed at the noise studies. Under the voltage exceeding operational one (that corresponds to accelerated reliability testing conditions), the check samples with SiO_2 films have shown pulse noise of telegraphic type in the oscillogram, which can be associated with processes of energizing-deenergizing of the surface conducting channels [51]. The defects occurring because of the presence of pores in SiO_2 films are probable cause of arising of the channels. In the batch with ZrO_2 protective films, only shot noise, which is in principle unavoidable, was observed. More detailed results of the device studies are presented in [52].

Thus, the studies performed on fianite and zirconium dioxide films as well as on the device structures developed using these films have demonstrated the advantages of zirconia-based solid solutions in application to photosensitive apparatus technology.

Studies of Optical Properties of ZrO_2 Films

Optical refraction of ZrO_2 equals to1.98÷2.1, that is, close to fianite one; therefore this material is also promising for antireflection coatings. Determination of the refraction constant n and monitoring of the film thickness d were carried out using ellipsometry technique. The experimentally determined values of d depended on duration of the films growth and varied within 600Å–1100Å range.

The films obtained have shown rather high refraction constant ~2÷2.1. These values were significantly higher than that of SiO_2 (1.45).

In theory, considering an incident beam from air (vacuum), it is possible to decrease the reflection to zero when the refraction constant of an antireflecting film corresponds to the following equation:

$$n = \sqrt{n_{f}}, \quad (1)$$

where -refraction constant of a semiconductor. In case of Si and GaAs, , thus . Therefore, ZrO_2 films obtained actually satisfy perfect antireflection for Si and GaAs-based devices from the viewpoint of . Moreover, the difference in -values of SiO_2 and ZrO_2 films provides an opportunity for the antireflection over a broad spectral range due to application of binary SiO_2 + ZrO_2 antireflecting coatings. The dependency of the reflection constant on wavelength of silicon sample coated with ZrO_2 film of 1200Å thickness is presented in Figure 18. Theoretical absorption minimum corresponds to μm.

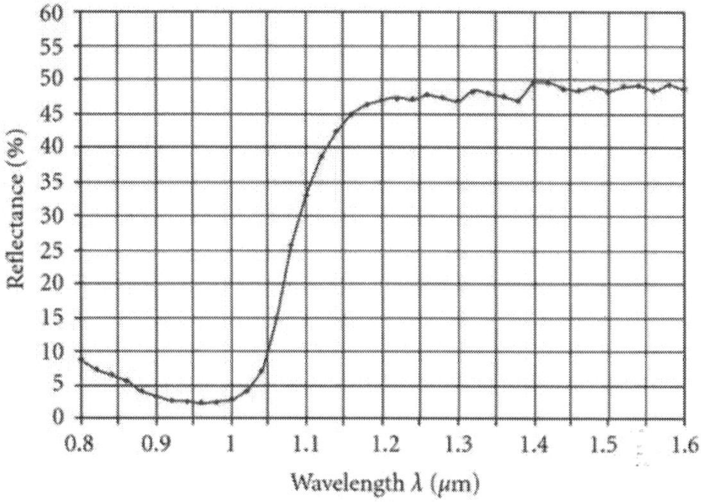

Figure 18. The dependency of the reflection on wavelength of silicon sample coated with ZrO_2 film of 1200Å thickness.

As it is apparent from Figure 18, the reflection minimum was approached at $\lambda_{min} = 0.97$ μm. Thus, the experimental results are in conformity with the theory practically complete.

Therefore, ZrO_2 film ensures high antireflection quality: at λ_{min}, the reflection loss does not exceed 2-3%. The data obtained confirm that ZrO_2 is an excellent material for antireflecting films, as well as fianite.

FIANITE AS ANTIREFLECTING LAYER FOR SOLAR CELLS

Antireflecting Properties of Fianite Film on Ge and Si

In theory, it is possible to eliminate the reflection completely (at the corresponding thickness of a film d) at, where n_f-optical refraction constant of a semiconductor. Since for Si and Ge the constants are equal to 3.7 and 4, respectively, the reflection is completely eliminated at $n_d = \sqrt{n_f} \approx 2$. Therefore, a dielectric having its optical refraction constant $n_d = \sqrt{n_f} \sim 2$ (at $n_f = 3.7 \div 4$) can be considered as an optimal material for the antireflecting film for solar cells and the other photosensitive devices. Theoretically, at the film thickness, which is equal to a quarter of optical wavelength $W = \lambda/4n_d$, such dielectric allows a complete elimination of the reflection loss (R=O).

The refraction constant of $SiO_2(n_d = 1.47)$ is considerably lower than that value. At this value it is impossible to maintain the reflection loss lower than 10%. Refraction constants of fianite and ZrO_2 are within (2.15÷ 2.18) and (2.13 ÷ 2.2), respectively, that is close to the above optimum value. Thus, providing an evidence that fianite and ZrO_2 are very promising as antireflecting coatings for solar cells and the other photosensitive devices based on Ge, Si, and $A^{III}B^{V}$compounds.

Experimental dependencies of antireflection (as dependencies of the reflection on wavelength) of fianite films on Si and Ge have been plotted (Figure 19).

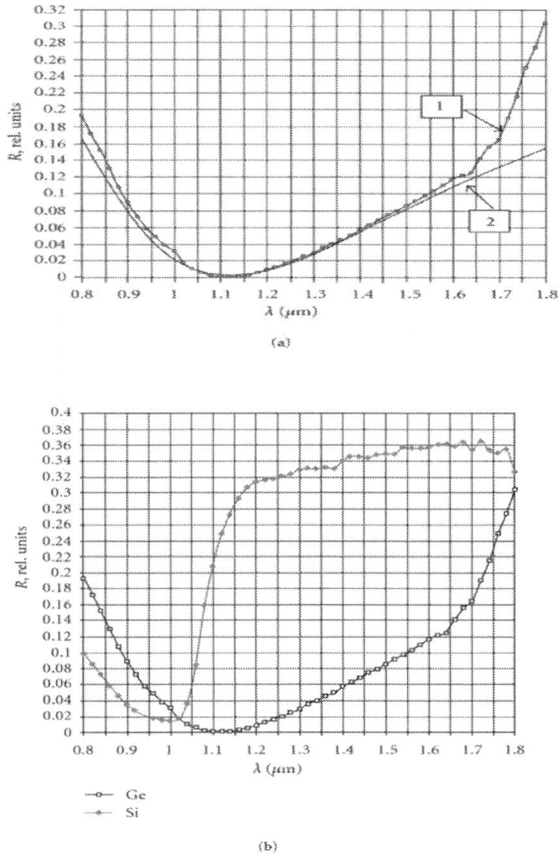

Figure 19. Experimental (1) and theoretical (2) dependencies of the reflection on wavelength in Ge-fianite antireflecting film system (1300Å) (a); experimental dependencies of the reflection of fianite film on Si and Ge (b).

The plots apparently demonstrate that the reflection drops to 0–1.5% in the minima.

Experimental study of antireflective properties of fianite oxide applied to Ge was performed. By the reason that germanium photodetectors are designed for detecting radiation generated by lasers with wavelengths $\lambda = 1.06,$ 1.3, and 1.54μm, thickness of the antireflective fianite film was chosen as W=1300Å; such thickness provides for minimal reflection losses in the considered wavelength range $\lambda = 1.06,$ –1.54μm. Figure 19(a) shows the comparison of experimental (thin line) and theoretical (bold line) $R(\lambda)$ curves. The theoretical $R(\lambda)$ curve was calculated using the following formula:

$$R = 1 - \frac{n_f n_d^2}{n_d^2(n_f+1)^2 - (n_f^2 - n_d^2)(n_d^2 - 1)\sin^2(2\pi n_d W/\lambda)}. \tag{2}$$

According to the above formula, reflection may fall practically to zero at the optimal value of n_d note that in case of SiO_2 antireflective film, for which n_d=1.47, it is impossible to obtain reflection lower than 10%). The minimal reflection is achieved at the following wavelength λ_{min}:

$$\lambda_{min} = 4Wn_d. \tag{3}$$

As it is shown in Figure 19(a), in the range of fundamental absorption (for λ<1.65 μm), the experimental curve 1 coincides with the theoretical curve 2. Some discrepancy at higher wavelengths (λ>1.65μm) appears due to deep penetration of such radiation and its reflection from the back surface. It is important that at the optimal wavelength (λ=1.12μm), fianite film provides for ideal antireflective properties, the reflection is actually absent. In rather large range 0.88–1.55 μm, into which radiation wavelengths of the most wide spread lasers fall, the losses for reflection do not exceed 10%.

The experimental dependences of enlightenment (the dependence of reflectance on the wavelength) of cubic zirconia films on Si and Ge (Ge by its optical properties is similar to GaAs) exhibit excellent antireflective properties of cubic zirconia (Figure 19(b)). As is evident from the graphs, the minimum reflection can drop to 0-1, 5%. Position of the minimum depends on the thickness of the film. When it gets thinned twice, the minimum would be in the solar spectrum. Plateau in the curve shows the reflection from the back side of the substrate in the transmission range for

Si. So the gain due to the use of the antireflecting fianite film reaches 20–30%.

So, it was experimentally proved that for 1300Å thick fianite film, reflection may actually drop to zero in the wavelength range λ=1.06–1.54 μm.

A new, unusual application of fianite as a reflecting film (in contrast to antireflecting!) was suggested. Such unexpected use may appear useful for screening of peripheral (nonphotosensitive) photodetector areas. Various routes of formation of photosensitive areas or metallic masks sputtered on SiO_2 are currently in use for the screening. But such solution causes notable spurious capacitance of the metal-oxide-semiconductor structure; provided that such capacities are inadmissible in a number of photodetectors, in particular, in high frequency photodetectors. In case of screening by the reflecting oxide (for this purpose, the thickness should be chosen as $W = 1/2\lambda n_d$), no surface capacity is being formed, because spurious capacitance is absent. In such case, fianite film may reflect about 60% of radiance from the surface.

Silicon Solar Cells with Fianite Antireflecting Layers

Experimental dependencies of antireflection of fianite films deposited on commercial solar cells were recorded. The reflection spectra of fianite obtained on two such samples are shown in Figure 20. The plots (Figures 20(a)and 20(b)) demonstrate excellent antireflecting properties of the fianite films. The plots also apparently demonstrate that the reflection drops to 0–1.5% in the minima. A position of the minimum depends on the film thickness. At the film thinning the minimum occurs in solar spectrum. Therefore, energy gain due to the application of the antireflecting fianite films approaches to 20–30%.

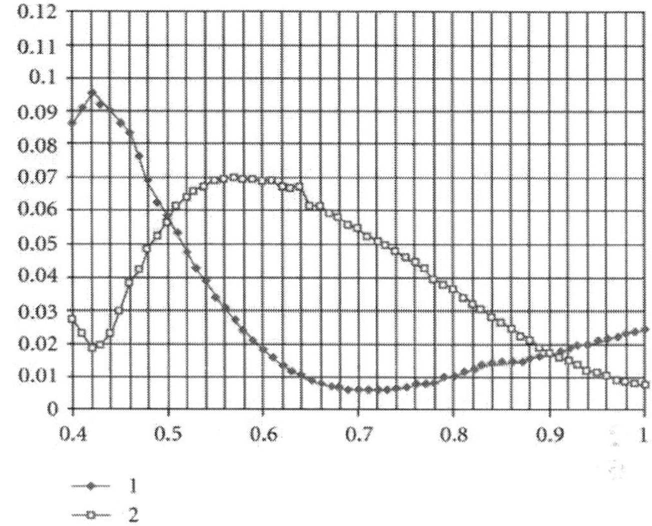
0.12
0.11
0.1
0.09
0.08
0.07
0.06
0.05
0.04
0.03
0.02
0.01
0
0.4
0.5
0.6
0.7
0.8
0.9
1
1
2

(a)

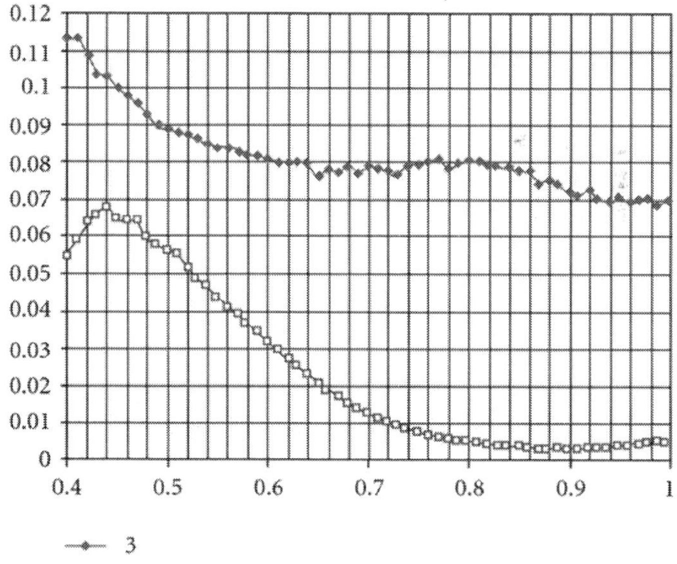
0.12
0.11
0.1
0.09
0.08
0.07
0.06
0.05
0.04
0.03
0.02
0.01
0
0.4
0.5
0.6
0.7
0.8
0.9
1
3
4

(b)

(c)

Figure 20. Antireflecting properties (reflection spectra) of fianite films of 580Å (a) and 1050Å (b) thickness. The experimental data were obtained on the industrial items s (c) of Si solar cells of "4 x 4" size.

CONCLUSIONS

Due to the unique combination of physical and chemical properties, fianite is a very promising multifunctional material for novel electronic technologies. Experimental data and some theoretical considerations presented in the review demonstrate prospects of the material application either as monolithic dielectric substrate or buffer layer for heteroepitaxy of semiconductor structures and as a material for insulating, antireflecting, and protective layers for various device structures, as well as a gate dielectric. Considering the performance characteristics, fianite (cubic zirconia) was shown to be in advance in comparison with the other materials currently in use. A number of specific application fields of the material in electronics were outlined.

REFERENCES

1. D. Pribat, L. M. Mercandalli, J. Siejka, and J. Perriere, "Interface oxidation of epitaxial silicon deposits on (100) yttria stabilized cubic zirconia," Journal of Applied Physics, vol. 58, no. 1, pp. 313–320, 1985.
2. E. I. Rau, A. N. Zhukov, and E. B. Yakimov, "Application of surface electron beam induced voltage method for the contactless characterization of semiconductor structures," Solid-State Phenomena, vol. 63-64, pp. 327–332, 1998.
3. G. G. Shahidi, "SOI technology for the GHz era," IBM Journal of Research and Development, vol. 46, no. 2-3, pp. 121–131, 2002.
4. G. G. Shahidi, "SOI technology for the GHz era," IBM Journal of Research and Development, vol. 46, no. 2-3, pp. 121–131, 2002.
5. G. Shengurov, V. N. Shabunov, A. N. Buzynin, et al., "Silicon heterostructures on fianite substrates,"Microelectronics, vol. 6, p. 204, 1996.
6. Golecki, H. M. Manasevit, L. A. Moudy, J. J. Yang, and J. E. Mee, "Heteroepitaxial Si films on yttria-stabilized, cubic zirconia substrates," Applied Physics Letters, vol. 42, no. 6, pp. 501–503, 1983.
7. H. Fukumoto, T. Imura, and Y. Osaka, "Heteroepitaxial growth of yttria-stabilized zirconia (YSZ) on silicon," Japanese Journal of Applied Physics, vol. 27, no. 8, pp. L1404–L1405, 1988.
8. H. I. Smith and D. C. Flanders, "Oriented crystal growth on amorphous substrates using artificial surface-relief gratings," Applied Physics Letters, vol. 32, no. 6, pp. 349–350, 1978
9. J. Steckl, J. Devrajan, C. Tran, and R. A. Stall, "SiC rapid thermal carbonization of the (111)Si semiconductor-on-insulator structure and subsequent metalorganic chemical vapor deposition of GaN,"Applied Physics Letters, vol. 69, no. 15, pp. 2264–2266, 1996.
10. L. M. Mercandalli, D. Diemegand, M. Crose, and Y. Sierka, "Recent progress in epitaxial growth of semiconducting materials on stabilized zirconia single crystals," Proceedings of the Society of Photo-Optical Instrumentation Engineers, vol. 623, pp. 133–210, 1986.
11. L. Wang, X. Liu, Y. Zan et al., "Wurtzite GaN epitaxial growth on a Si(001) substrate using γ-Al_2O_3 as an intermediate layer," Applied Physics Letters, vol. 72, no. 1, pp. 109–111, 1998.
12. M. F. Wu, C. Chen, D. Zhu et al., "Depth dependence of the tetragonal distortion of a GaN layer on Si(111) studied by Rutherford backscattering/channeling," Applied Physics Letters, vol. 80, no. 22, pp. 4130–4132, 2002.
13. M. Sumiya, Y. Kurumasa, K. Ohtsuka, K. Kuwahara, Y. Takano, and S. Fuke, "Reduction of defect density in GaN epilayer having buried Ga metal

by MOCVD," Journal of Crystal Growth, vol. 237–239, no. 1–4, pp. 1060–1064, 2002.

14. N. Buzynin and V. P. Kalinushkin Abstracts of the, "New technique of laser characterization of sapphire and fianite crystals," in Proceedings of the Laser Interaction with Matter International Symposium (LIMIS '10), Changchun, China, August 2010.
15. N. Buzynin and V. P. Kalinushkin, "Laser characterization of sapphire defectts," in Proceedings of the International Congress " Materials Week-2001" on Advanced Materials, their Processes and Applications, article 389, Munich, Germany, October 2001.
16. N. Buzynin, T. N. Grishina, T. V. Kiselyov et al., "Zirconia-based solid solutions-New materials of photoelectronics," Optical Memory and Neural Networks, vol. 18, no. 4, pp. 312–321, 2009.
17. N. Buzynin, V. V. Osiko, E. E. Lomonova, et al., "Epitaxial films of III-V compound on fianite substrates," in Proceedings of the International Congress on Advanced Materials, Their Processes and Applications, article 672, pp. 25–28, Munich, Germany, September 2000.
18. N. Buzynin, V. V. Osiko, E. E. Lomonova, N. Yu. Buzynin, and A. S. Usikov, "Epitaxial films of GaAs and GaN on fianite substrate," in Proceedings of the Materials Research Society Symposium, vol. 512 of Wide-Bandgap Semiconductors for High Power, High Frequency and High Temperature, pp. 205–210, Pittsburg, Pa, USA, 1998.
19. N. Buzynin, V. V. Osiko, K. Yu. Voron'ko, et al., "Epitaxial GaN and GaNxAs1−x films on monolithic and porous GaAs substrates with fianit sublayer," Bulletin of the Russian Academy of Sciences: Physics, vol. 69, no. 4, pp. 557–562, 2005.
20. N. Buzynin, V. V. Osiko, Y. K. Voronko et al., "Epitaxial fianit films on Si and GaAs," Bulletin of the Russian Academy of Sciences. Physics, vol. 67, no. 4, pp. 586–587, 2003.
21. N. Buzynin, V. V. Osiko, Y. K. Voronko et al., "Epitaxial structures of AIIIBV materials on fianite,"Izvestiya Akademii Nauk. Seriya Fizicheskaya, vol. 66, no. 9, pp. 1345–1351, 2002.
22. N. Buzynin, V. V. Osiko, Y. N. Buzynin et al., "Fianite: a multipurpose electronics material," Bulletin of the Russian Academy of Sciences: Physics, vol. 74, no. 7, pp. 1027–1033, 2010.
23. N. Buzynin, V. V. Osiko, Y. N. Buzynin, and B. Pushnyi, "Growth on GaN and GaAs on fianite by MOCVD capillary epitaxy technique," MRS Internet Journal of Nitride Semiconductor Research, vol. 4, Article 49, 1998.
24. N. Buzynin, Y. N. Buzynin, A. V. Belyaev, A. E. Luk'yanov, and E. I. Rau, "Growth and defects of GaAs and InGaAs films on porous GaAs substrates," Thin Solid Films, vol. 515, no. 10, pp. 4445–4449, 2007.

25. N. N. Sheftal and A. N. Buzynin, "Predominant orientation of crystals on a substrate and the effect of scratches," Vestnik Moskovskogo Universiteta, no. 3, series 4, pp. 102–104, 1972.
26. N. P. Kobayashi, J. T. Kobayashi, P. D. Dapkus, et al., "GaN growth on Si(111) substrate using oxidized AlAs as an intermediate layer," Applied Physics Letters, vol. 71, Article ID 3569, 3 pages, 1997.
27. N. Pryds, B. Toftmann, J. B. Bilde-Sørensen, J. Schou, and S. Linderoth, "Thickness determination of large-area films of yttria-stabilized zirconia produced by pulsed laser deposition," Applied Surface Science, vol. 252, no. 13, pp. 4882–4885, 2006.
28. N. Wakiya, T. Yamada, K. Shinozaki, and N. Mizutani, "Heteroepitaxial growth of CeO_2 thin film on Si(001) with an ultra thin YSZ buffer layer," Thin Solid Films, vol. 371, no. 1, pp. 211–217, 2000.
29. Ohtani, K. S. Stevens, and R. Beresford, "Microstructure and photoluminescence of GaN grown on Si(111) by plasma-assisted molecular beam epitaxy," Applied Physics Letters, vol. 65, no. 1, pp. 61–63, 1994.
30. Osinsky, S. Gangopadhyay, J. W. Yang et al., "Visible-blind GaN Schottky barrier detectors grown on Si(111)," Applied Physics Letters, vol. 72, no. 5, pp. 551–553, 1998.
31. P. A. Anderson, C. E. Kendrick, R. J. Kinsey et al., "(111) and (100) YSZ as substrates for indium nitride growth," Physica Status Solidi C, vol. 2, no. 7, pp. 2320–2323, 2005
32. P. Caricato, G. Barucca, A. Di Cristoforo et al., "Excimer pulsed laser deposition and annealing of YSZ nanometric films on Si substrates," Applied Surface Science, vol. 248, no. 1–4, pp. 270–275, 2005
33. P. D. C. King, T. D. Veal, S. A. Hatfield et al., "X-ray photoemission spectroscopy determination of the InN/yttria stabilized cubic-zirconia valence band offset," Applied Physics Letters, vol. 91, no. 11, Article ID 112103, 3 pages, 2007.
34. R. Lyonnet, A. Khodan, A. Barthélémy et al., "Pulsed laser deposition of $Zr_{1-x}Ce_xO_2$ and $Ce_{1-x}La_xO_{2-x/2}$for buffer layers and insulating barrier in oxide heterostructures," Journal of Electroceramics, vol. 4, no. 2, pp. 369–377, 2000.
35. R. Paszkiewicz, B. Paszkiewicz, R. Korbutowicz, et al., "MOVPE GaN grown on alternative substrates,"Crystal Research and Technology, vol. 36, no. 8–10, pp. 971–977, 2001.
36. S. Abermann, G. Pozzovivo, J. Kuzmik et al., "MOCVD of HfO_2 and ZrO_2 high-k gate dielectrics for InAlN/AlN/GaN MOS-HEMTs," Semiconductor Science and Technology, vol. 22, no. 12, pp. 1272–1275, 2007.
37. S. Guha and N. A. Bojarczuk, "Ultraviolet and violet GaN light emitting diodes on silicon," Applied Physics Letters, vol. 72, no. 4, pp. 415–417, 1998.

38. S. J. Wang, C. K. Ong, S. Y. Xu et al., "Crystalline zirconia oxide on silicon as alternative gate dielectrics,"Applied Physics Letters, vol. 78, no. 11, pp. 1604–1606, 2001.
39. S. J. Wang, C. K. Ong, S. Y. Xu et al., "Electrical properties of crystalline YSZ films on silicon as alternative gate dielectrics," Semiconductor Science and Technology, vol. 16, no. 3, pp. L13–L16, 2001.
40. S. Jeon, M. Takanori, A. Unno, K. Wasa, Y. Ichikawa, and H. Hwang, "Interfacial properties of a hetero-structure YSZ/p-(1 0 0)Si prepared by magnetron sputtering," Vacuum, vol. 65, no. 1, pp. 19–25, 2002.
41. S. Kaneko, K. Akiyama, T. Ito et al., "Single domain epitaxial growth of yttria-stabilized zirconia on Si(1 1 1) substrate," Ceramics International, vol. 34, no. 4, pp. 1047–1050, 2008.
42. S. Yu. Kuz'minov, E. E. Lomonova, and V. V. Osiko, Cubic Zirconia and Skull Melting, Cambridge International Science, UK, 2009.
43. S.-C. Hwang and H. S. Shin, "Effect of deposition temperature on the growth of yttria-stabilized zirconia thin films on Si(111) by chemical vapor deposition," Journal of the American Ceramic Society, vol. 82, no. 10, pp. 2913–2915, 1999.
44. T. Hata, K. Sasaki, Y. Ichikawa, and K. Sasaki, "Yttria-stabilized zirconia (YSZ) heteroepitaxially grown on Si substrates by reactive sputtering," Vacuum, vol. 59, no. 2-3, pp. 381–389, 2000.
45. T. Nakamura, Y. Tokumoto, R. Katayama, T. Yamamoto, and K. Onabe, "RF-MBE growth and structural characterization of cubic InN films on yttria-stabilized zirconia (0 0 1) substrates," Journal of Crystal Growth, vol. 301-302, pp. 508–512, 2007.
46. T. Ngai, W. J. Qi, R. Sharma et al., "Electrical properties of ZrO_2 gate dielectric on SiGe," Applied Physics Letters, vol. 76, no. 4, pp. 502–504, 2000.
47. V. G. Beshenkov, A. G. Znamenskii, V. A. Marchenko, A. N. Pustovit, and A. V. Chernykh, "Widening temperature range of epitaxial growth of YSZ films on Si [100] under magnetron sputtering," Journal of Technical Physics, vol. 77, no. 5, pp. 102–107, 2007.
48. V. I. Aleksandrov, V. V. Osiko, A. M. Prokhorov, and V. M. Tatarintsev, "New technique of the synthesis of refractory single crystals and molten ceramic materials," Vestnik Akademii Nauk SSSR, no. 12, pp. 29–39, 1973 (Russian).
49. V. V. Osiko, M. A. Borik, and E. E. Lomonova, "Synthesis of refractory materials by skull melting technique," in Handbook of Crystal Growth, pp. 433–469, Springer, 2010.

CITATION

Alexander Vodopyanov, Yurii Buzynin, Dmitry Mansfeld, Oleg Khrykin, Yurii Drozdov, Pavel Yunin, Andrei Lukyanov, Mikhail Viktorov, Sergey Golubev, and Vladimir Shashkin, "Monocrystalline InN Films Grown at High Rate by Organometallic Vapor Phase Epitaxy with Nitrogen Plasma Activation Supported by Gyrotron Radiation," Japanese Journal of Applied Physics, vol. 52, no. 11, 2013.

CHAPTER 4

CMOS Compatible Bulk Micromachining

John Ojur Dennis1, Farooq Ahmad1 and M. Haris Khir1

[1] Department of Fundamental and Applied Sciences, Department of Electrical and Electronic Engineering, Universiti Teknologi PETRONAS, Perak, Malaysia

INTRODUCTION

The last two decades have seen the emergence and prevalence of Micro-Electro-Mechanical Systems (MEMS) technology. The reduction in size, low power consumption and low cost is the ultimate goals of application specific MEMS devices. One way to achieve these goals is the monolithic integration of MEMS technology with the standard integrated circuit (IC) technology. The utilization of the conventional IC technology as a platform for design of the fast-growing MEMS technology has led to development of numerous devices and technologies in the past years. Much effort has been made and large capital has been invested into mainstream Complementary Metal-Oxide Semiconductor (CMOS) compatible MEMS technology. This approach has enabled MEMS devices to be directly integrated with CMOS circuits, allowing smaller device sizes and higher performances. This integration of MEMS technology with the mainstream CMOS technology is referred to as CMOS-MEMS technology [1].

BULK MICROMACHINING TECHNOLOGIES

Among all micromachining technologies, Bulk micromachining is the oldest one. To make small mechanical components using bulk micromachining technology, substrate material such as silicon wafer is selectively etched. There are two major classes of bulk micromachining: wet bulk micromachining (WBM) and dry bulk micromachining (DBM).WBM gives nice feature resolution and vertical sidewalls for deep trenches in the substrate but have numerous disadvantages such as incompatibility with microelectronic circuits and higher cost due to mask requirement. The fabrication cost, time and simplification of micromachining for MEMS devices is reduced by using dedicated MEMS foundry and making the prototyping development cycle much shorter by reducing number of masks and photolithography steps. CMOS MEMS technology is a good solution for post micromachining because in this technology CMOS circuitry is protected completely by the top metal layer and the same top metal layer is used as mask to define MEMS structures in MEMS region during the post-CMOS micromachining. That is why this post micromachining process is known as CMOS-compatible.

Wet Bulk Micromachining

Wet bulk micromachining is done normally with the help of chemicals. In chemical wet etching, the silicon substrate is immersed into the solution of reactive chemicals which etches the exposed regions of substrate at measureable rates. Wet etching typically uses alkaline liquid solvents, such as potassium hydroxide (KOH) or tetramethylammonium hydroxide (TMAH) to dissolve silicon which has been left exposed by the photolithography masking step. These alkali solvents dissolve the silicon in a highly anisotropic way, with some crystallographic orientations dissolving up to 1000 times faster than others. Fig. 1 shows the three major crystal planes in a cubic unit of silicon.

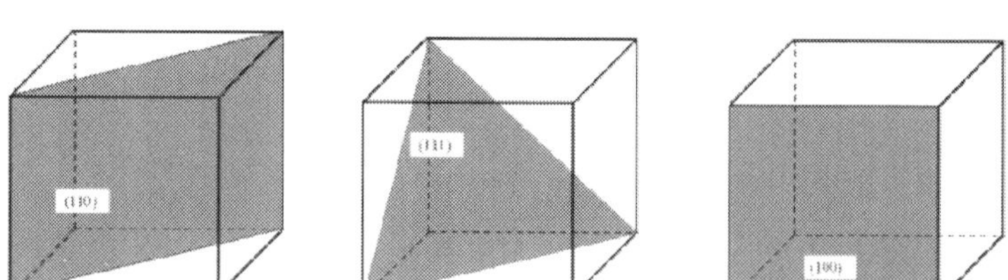

Figure 1. The three orientations <110>, <111> and <100> are the respective directions normal to the plane.

For example in silicon anisotropic etching, planes <111> are etched at slower rate than all other planes. The reasons for the slow etch rate of planes <111> are, high density of silicon atoms exposed to the etchant solution in this direction and three silicon bonds laying below the plane. Fig. 2 (a) shows a schematic and (b) three dimensional photo of a typical wet anisotropic etching of a silicon substrate.

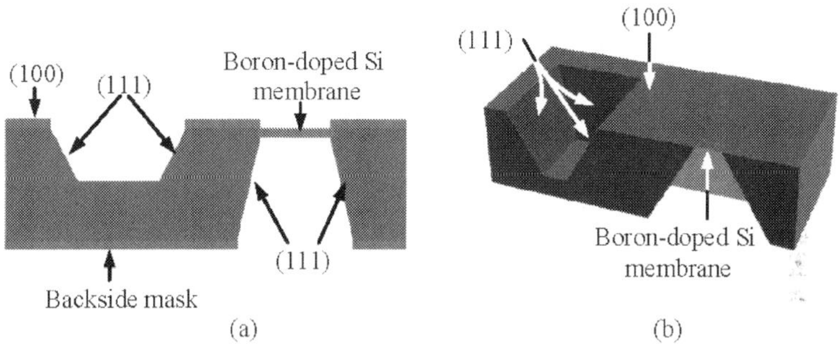

Figure 2. (a) 2-D Schematic and (b) 3-D diagram of a typical wet bulk micromachining of Si.

Wet chemical micromachining technology is popular in MEMS industry because of high etch rate and selectivity. There is one drawback and that is the mask is also etched during normal etching process so it is suggested to find a mask that doesn't dissolve or at least dissolves at a much slower rate than the silicon substrate. In wet etching, etch rates and selectivity can be modified by various methods such as: (a) chemical composition of etching solution, (b) dopant concentration in the substrate, (c) adjusting temperature of etching solution and finally (d) Modifying crystallographic planes of the substrate.

Chemical wet etching in bulk micromachining may be further subdivided into two parts: isotropic wet etching where the etch rate is not dependent on crystallographic orientation of the substrate and etching proceeds in all directions at equal rates as shown in Fig.3 (a) and anisotropic wet etching where the etch rate is dependent on crystallographic orientation of substrate as shown in Fig. 3 (b). Often in order to control etching process and uniform etch depths across the wafer, etch stops are used. Generally three types of etch stops are used in micromachining: Dopant etch stops, Electrochemical etch stops and Dielectric etch stop.

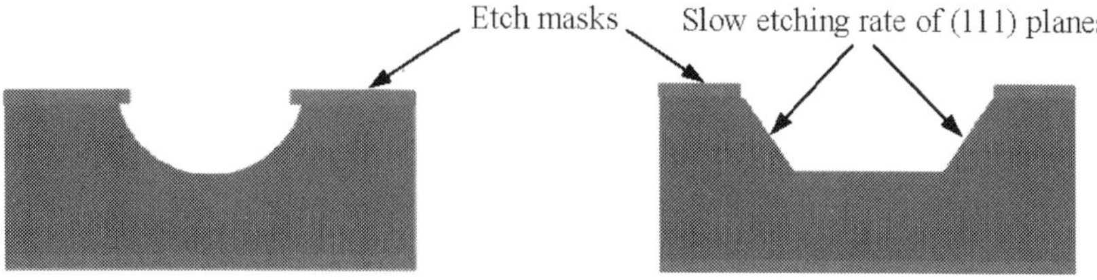

(a) Isotropic (b) Anisotropic

Figure 3. Difference between (a) isotropic and (b) anisotropic wet etching.

Usually in order to control etching process and uniform etch depths across the wafer, etch stops are used. Generally three types of etch stops are used in micromachining: Dopant etch stops, Electrochemical etch stops and Dielectric etch stop [2].

Dry Bulk Micromachining

Dry etching in bulk micromachining is classified into three groups: Reactive ion etching (RIE), Vapor phase etching and Sputter etching. Reactive ion etching (RIE) uses both physical and chemical mechanisms to achieve high levels of resolution. The process is one of the most diverse and most widely used processes in industry and research. Since the process combines both physical and chemical interactions, the process is much faster. The high energy collision from the ionization helps to dissociate the etchant molecules into more reactive species. In the RIE-process, cations are produced from reactive gases which are accelerated with high energy to the substrate and chemically react with the silicon. The typical RIE gasses for Si are CF_4, SF_6 and $BCl_2 + Cl_2$.

Both physical and chemical reaction is taking place with the physical part similar to the sputtering deposition. If the ions have high enough energy, they can knock atoms out of the material to be etched without a chemical reaction. It is very complex task to develop dry etch processes that balance chemical and physical etching, since there are many parameters to adjust. By changing the balance it is possible to influence the anisotropy of the etching with the chemical part being isotropic and the physical part highly anisotropic and therefore the combination can form side walls that have shapes from rounded to vertical.

A special subclass of RIE which continues to grow rapidly in popularity is deep RIE (DRIE). In this process, etch depths of hundreds of microns can be achieved with almost vertical sidewalls. The primary technology is based on the so-called "Bosch process", named after the German company Robert Bosch which filed the original patent, where two different gas compositions are alternated in the reactor. The first gas composition etches the substrate, and the second gas composition creates a polymer on the surface of the substrate. The polymer is immediately sputtered away by the physical part of the etching, but only on the horizontal surfaces and not the sidewalls. Since the polymer only dissolves very slowly in the chemical part of the etching, it builds up on the sidewalls and protects them from etching and working as pasivation, as illustrated in fig. 4. In fig.4 (a) SF_6 is etching the silicon while in fig. 4 (b) C_4F_8 is playing the role of passivation and in fig. 4 (c) again SF_6 is etching. As a result, etching aspect ratios of 1 to 50 can be achieved. The process can easily be used to etch completely through a silicon substrate, and etch rates are 3-4 times higher than wet etching. In the following part more details are given on DRIE. Under our MEMS research group in Universiti Teknologi PETRONAS, high aspect ratio of 3 to 50 μm and complete etch through is achieved as shown in fig.5.

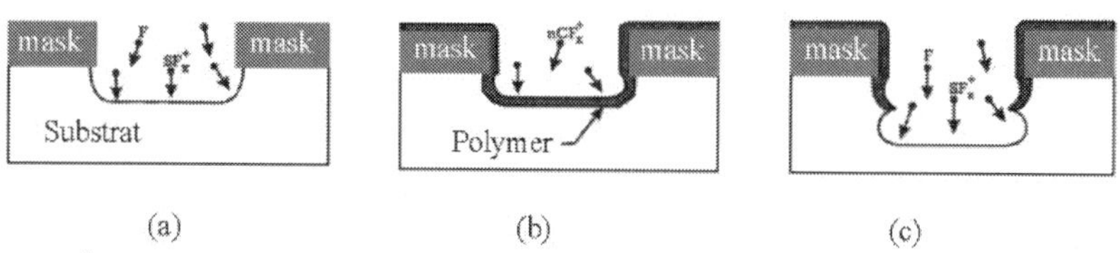

Figure 4. Illustration of etching and passivation in DRIE.

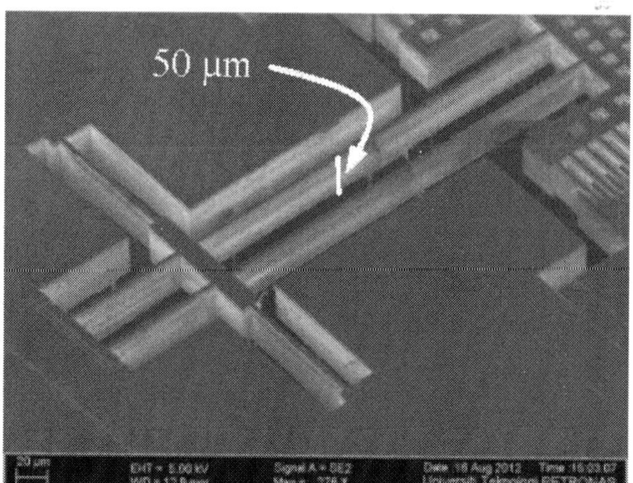

Figure 5. FESEM image of a high aspect ratio plasma etching using DRIE with inhibitor.

Vapor phase etching (also called Chemical dry etching) does not use liquid chemicals or etchants. This process involves a chemical reaction between etchant gases to attack the silicon surface. The chemical dry etching process is usually isotropic and exhibits high selectively. Anisotropic dry etching has the ability to etch with finer resolution and higher aspect ratio than isotropic etching. Due to the directional nature of dry etching, undercutting can be avoided. Some of the ions that are used in chemical dry etching is tetrafluoromethane (CH_4), sulfur hexafluoride (SF_6), nitrogen trifluoride (NF_3), chlorine gas (Cl_2), or fluorine (F_2) [3]. Vapor phase etching can be done with simpler equipment than what RIE requires. The two most common vapor phase etching technologies are silicon dioxide etching using hydrogen fluoride (HF) and silicon etching using xenon diflouride (XeF_2), both of which are isotropic in nature. Usually, care must be taken in the design of a vapor phase process to not have bi-products formed in the chemical reaction that condense on the surface and interfere with the etching process.

Sputter etching is essentially RIE without reactive ions. The systems used are very similar in principle to sputtering deposition systems. The big difference is that the substrate is now subjected to ion bombardment instead of the material target used in sputter deposition [4].

CMOS COMPATIBLE MICROMACHINING

CMOS compatible micromachining is now commonly known as CMOS MEMS technology. During post processing in CMOS MEMS technology, the top metal layer acts as mask to define MEMS structures. For the front side processing of the die no photolithographic steps are required. CMOS MEMS technology facilitates the wiring of MEMS structures to integrated circuits due to many metal layers. Through monolithic integration of the CMOS-MEMS devices with circuits, the parasitic capacitances is reduced and overall high performance devices is achieved. Dry post CMOS MEMS technologies are divided into two classes: Thin film micromachining technology (Thin-film Post-CMOS MEMS Technology) and DRIE bulk micromachining technology (DRIE Post-CMOS MEMS Technology).

Thin-Film Post-CMOS MEMS Technology

The process flow of thin-film CMOS-MEMS with three metal layers is shown in Fig. 6. The CMOS circuit region is designed to be covered by the top metal layer and the MEMS structures are also pre-defined by the top metal layer or the other interconnect metal layers. Fig.6 (a) shows the cross-section of the original chip after CMOS foundry fabrication, with a passivation layer on top. Two processing steps are performed only on the front side of the chip. First, the pre-defined MEMS structure is opened by etching the SiO_2 dielectric stack layer as shown in Fig.6 (b). This is done by an anisotropic SiO_2plasma etch using Tegal SS110 A DRIE system based on Bosch technology [5]. Next, an anisotropic silicon DRIE is performed using the same Tegal DRIE system as shown in Fig.6 (c), followed by isotropic silicon etch, which releases the MEMS structure by undercutting the silicon beneath as shown in Fig.6 (d). The depth of the anisotropic etch into the silicon controls the gap between the released structure and the silicon substrate. This gap should be large enough to eliminate the parasitic capacitance between the MEMS structures and the silicon substrate. In practice, it is normally on the order of 30 μm [6].

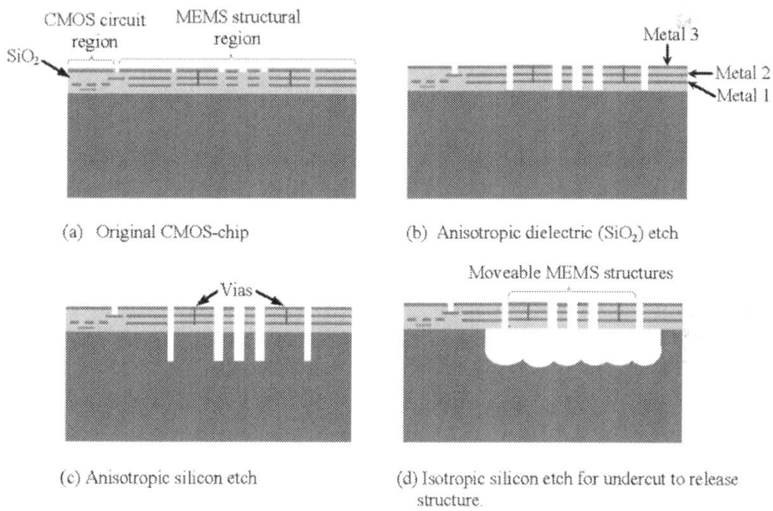

Figure 6. The process-flow for thin film DRIE post-CMOS micromachining.

A resonant magnetic field sensor and piezoresistive accelerometer are examples of devices fabricated using this technology [7, 8]. This simple fabrication process yields much smaller parasitic effects as compared to the other processes such as poly MUMPs and it provides flexible wiring by using the multiple metal layers. However, there are some drawbacks in this technology, which limit the performance of the fabricated accelerometers. First, there is large vertical curling in the comb fingers of the suspended MEMS structures caused by the residual stress gradient existing in the composite SiO_2/Al layers as shown in Fig. 7. Additionally, due to different thermal expansion coefficients (TEC) of Al and SiO_2, the curled MEMS devices exhibit a strong temperature dependence, which limits their utility. Although particular compensation technology was employed to reduce the structure mismatch by using a specially designed frame, the device fabricated using this thin film process still has a stringent size limit.

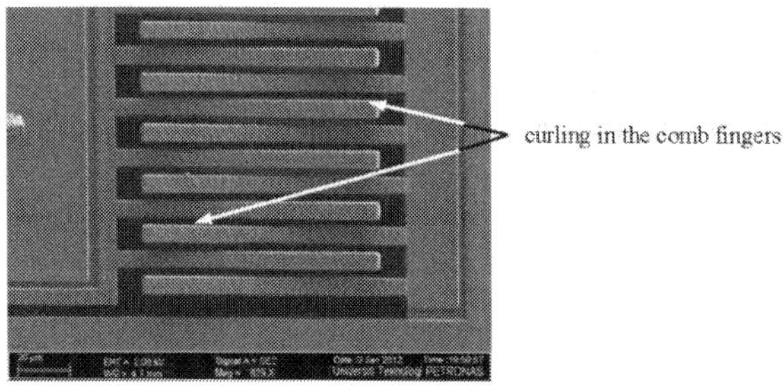

Figure 7. FESEM image of typical comb fingers fabricated using thin film CMOS-MEMS technology.

Second, for sensors which require proof mass like inertial sensors fabricated using this technology, the requirement of holes on the proof mass reduces the proof mass, resulting in a lower mechanical sensitivity of the sensors. To achieve a capacitance change large enough for the readout circuit to detect, the dimension of the sensors may need to be considerably large, which is in conflict with the dimensional limit imposed by the possibility of structure curling. Third, in-plane curling of the thin film structures due to fabrication variations also limits the maximum size of the device.

Although there are some specific processes designed for low residual stress thin films [9], Farooq et al utilized the Malaysian Institute of Microelectronic Systems (MIMOS Bhd) 0.35 μm CMOS foundry processes to reduce the curling of the MEMS structures by proper design of the structures as described in [10]. Fig. 8 shows the successfully released central shuttle of magnetic field sensor fabricated using this process. It can be clearly observed that the curling in the structure has been significantly reduced as indicated by the minor curling in the fingers. In addition to the collection of technological data directly from the employed CMOS foundries, systematic post-CMOS process calibration must be conducted to characterize the process variations and their effects on the MEMS devices.

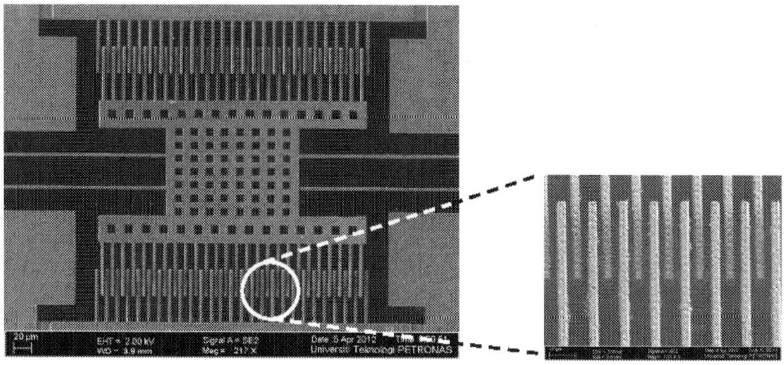

Figure 8. FESEM image showing comb fingers of the central part of a magnetic field sensor.

Lastly, since the last release step is an isotropic etch, it undercuts the silicon close to the circuit and structure anchors. The ratio of the vertical and lateral etching rate is approximately 3:5. To protect the silicon underneath the circuit region from being etched away during the structure release, the CMOS circuitry must be placed far away from the microstructures, especially when a large separation between the microstructures and substrate is needed. As a result, significant chip area is wasted due to the protection margin around the MEMS structures. Since the silicon underneath the mechanical anchors of the MEMS structures is also etched away, the suspension of the mechanical structures is softened, which results in a lower mechanical performance and less robustness of the device [10].

DRIE Post-CMOS MEMS Technology

To reduce the disadvantages of the above thin-film post-CMOS MEMS process, DRIE post-CMOS MEMS technology is developed to incorporate bulk single crystal silicon (SCS) into the MEMS structures that are formed by CMOS thin films. Due to the improvement in DRIE technology, high aspect ratio CMOS MEMS structures have been exhibited as shown in Fig. 9. The achievement of maximum aspect-ratio in DRIE system is an important factor in the MEMS structure design [5]. It decides the dimensional limit of the structures fabricated using that DRIE system. Once a lateral feature to be etched is fixed, the maximum thickness is uniquely defined.

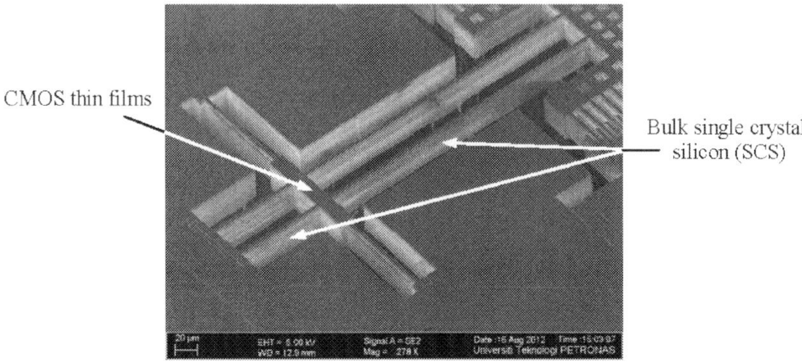

Figure 9. FESEM image showing the incorporation of bulk SCS to the CMOS thin films.

Process Flow

The DRIE post-CMOS MEMS technology branches from the thin film technology described in the previous section. DRIE CMOS-MEMS provides an approach to implement thick and flat MEMS structures with improved mechanical performance and device strength due to advantage of both flexible wiring with multiple metal layers in CMOS technology and the capability of high aspect ratio etching.Fig.10 shows the process flow of the DRIE post-CMOS MEMS technology.

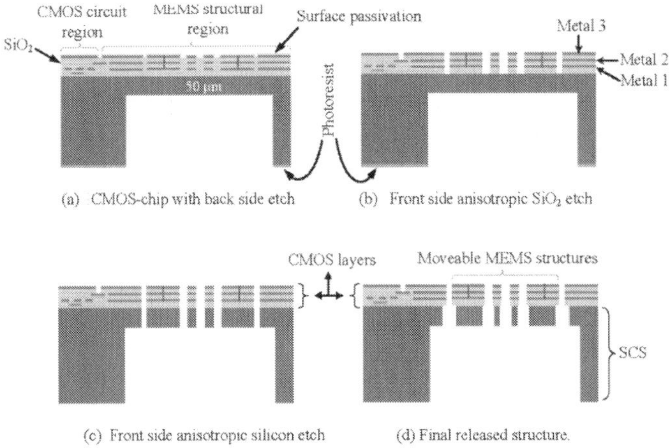

Figure 10. The process-flow for DRIE CMOS micromachining.

To define the MEMS structure thickness (thin film plus substrate), the process starts with backside silicon DRIE etching by the selective application of photoresist at the back-side as shown in Fig.10 (a). The maximum thickness of this structure is limited by the smallest etching pattern on the front side of the MEMS structure and the maximum etch aspect-ratio. Next, to define the MEMS structures, an anisotropic SiO_2 etch is performed on the front side of the chip as shown in Fig. 10 (b) similar to the thin film process. In the following step, the structure is released by etching through the remaining SCS substrate. This step differs from the thin film process in that an anisotropic DRIE, instead of isotropic Si etch, is performed to release the structure as shown in Fig.10 (c). Flat and large MEMS structures can be obtained by incorporating SCS underneath the CMOS interconnect layers because the residual stresses in the SiO_2/Al thin-films are mitigated by the thick SCS, showing very little out-of-plane curling as shown in Fig. 11 where it can be seen that the sense fingers are perfectly flat and aligned.

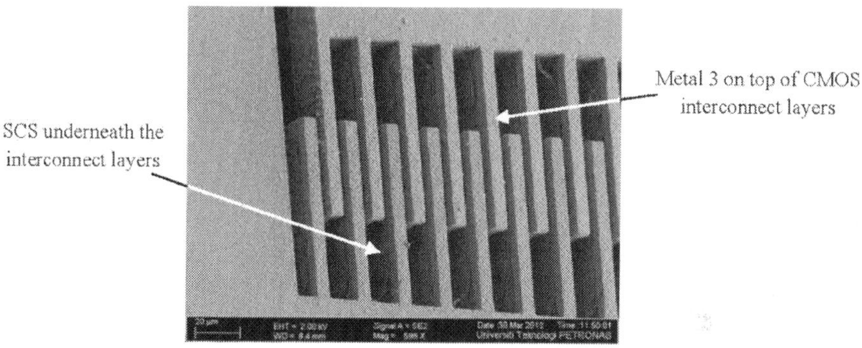

Figure 11. FESEM image showing comb fingers with thick SCS underneath.

In some MEMS structures isotropic silicon etch can be added to create nice mechanical structures which only consist of CMOS thin films. A resonant magnetic field sensor fabricated under University Technology PETRONAS (UTP) MEMS Research group incorporates both CMOS thin-film layers for the beams and silicon substrate along with CMOS thin films in the side torsional coils as shown in Fig. 12.

Figure 12. FESEM image of CMOS thin layers and incorporation of SCS with thin layers as MEMS structures.

Double side alignment is required in the backside DRIE step to define the thickness of the MEMS structures. It is observed that normally the remaining silicon for bulk MEMS structures is on the order of tens of microns and by the substrate thinning there is no apparent circuit performance degradation. Therefore, there is no hard and fast constraints for alignment in the backside etch. A reliable sensing and actuation is achieved with the flat MEMS microstructures enabled by the DRIE CMOS MEMS technology. A resonant magnetic field sensor with piezoresistive sensing is fabricated using this technology [11]. The technology is very suitable for the fabrication of, in particular, thermally actuated micromirrors where bimorphs are used to elevate the mirror plate. Numerous resonant chemical sensors, electrothermal micromirrors and accelerometers have been fabricated using this technology [12].

Design Rules for MEMS Devices

For CMOS circuit designs, there are different ways to break up a layout into basic elements. Designers usually divide and analyze a circuit by the function of individual components, devices and interconnect. The devices may be active (e.g. transistors, diodes) or passive devices (e.g. resistors, capacitors and inductors). The interconnects include in-plane wires, pads and vias which connect multiple metal layers. The critical information on the geometric patterns includes the size of each feature, the spacing between features and the overlap of features. These geometric data, required in the design, leads to evolutions of processes and ensures successful fabrication of microelectronic circuitry. In order to fulfill the requirements of successful MEMS device designs, processing information, which quantitatively describes the capabilities and limitations of the process under predetermined processing conditions and parameter settings, is necessary and crucial. This information is abstracted as MEMS design rules and is verified in the design environment as illustrated in Fig. 13.

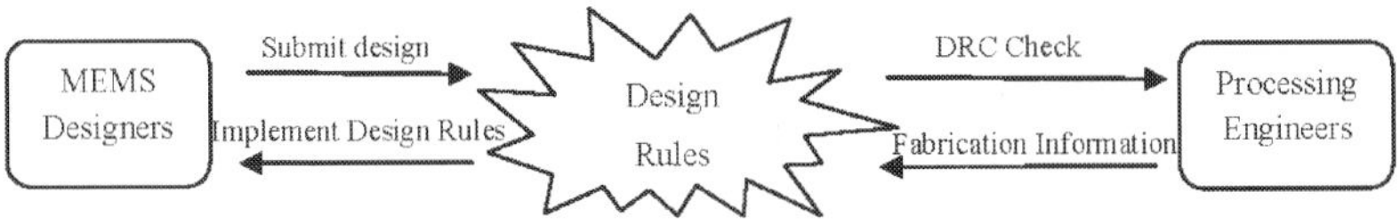

Figure 13. Design rules function between MEMS designers and MEMS fabrication engineers.

This section explains how to construct concise and conclusive design rules. We used MIMOS Berhad Malaysia standard 0.35 μm CMOS design rules for the fabrication of CMOS MEMS devices and only the design rules for mechanical structures are discussed in this section. These MEMS design rules add further constraints to layouts. Similar to CMOS design rules, a complex micromechanical layout is fragmented into small pieces, and MEMS design rules contain only fundamental geometries directly related to the post CMOS micromachining process. A set of geometric constraints and design rules are then based on the final process.
Design rules are expressed in the design-rule file used in layout tools such as Cadence Virtuoso, Magicor, Mentor Graphic, layout tool etc. After all processing completion, a post-CMOS micromachined chip can be partitioned into a released area (containing free standing parts and free moving parts) and a non-released area (anchored parts) according to the mechanical behavior. The objective of post CMOS processing is the final result of releasing microstructures and maintaining critical dimensions. There are practical difficulties in monitoring each intermediate process stages. Therefore characterization of the process to extract design rules is performed at the end of the entire process flow. An overview of a typical post CMOS MEMS magnetic sensor is shown in Fig. 14.

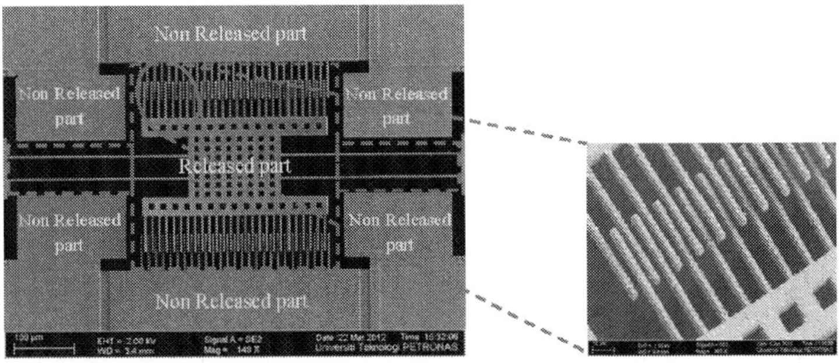

Figure 14. Partitioning of a MEMS sensor for design-rule checking.

The anchored part is the area outside of the red dash-line box on FESEM, whereas the released part is the area inside the red border. The close-up FESEM view of circled portion is shown in Fig. 14 (b). Sufficient paths for the gas flow are required in order to allow the reactive radicals in the plasma to undercut MEMS structures. These paths can surround mechanical structures, so that the undercut can happen from the periphery of the structures. The gas flow can also pass through structures via etch release holes. Fig. 14 illustrates various situations for reactive species undercut: from one side of the structures toward the anchor region, from both sides of beam structures, or outward through etch release holes on plates.

Theoretically, any shape of the feature which opens a gas flow path from the top of the chip to the substrate is suitable for the purpose of undercut. However, currently in CMOS foundries, Manhattan layout is preferred, or at least convenient. The majority of CMOS foundries only accept patterns with edges which are parallel, perpendicular or 45° with respect to the horizontal/vertical alignment lines. Non-Manhattan features require more and smaller rectangular patterns to approximate the geometry, and therefore, require a larger data base and longer exposure time, which leads to higher cost. Today, raster-scanning mask-making technologies such as e-beam or laser systems can overcome the difficulties mentioned above but non-Manhattan shapes approximated by a smaller raster-beam size still require more time in fabrication [13]. Following the practice of CMOS fabrication, this work only explores rectangular design rules. The major purpose of this section is to ensure the release of moving/free-standing parts and to guarantee the integration of anchor parts. Other process related rules are not included. According to situations described in Fig. 14, there have been three features encountered in the MEMS structure layout for the purpose of undercut: large openings, gaps and holes. To control the etch rates in these opening, dummy structures are used as shown in Fig.15 (a) whilefigure 15 (b) shows the fully released structure.

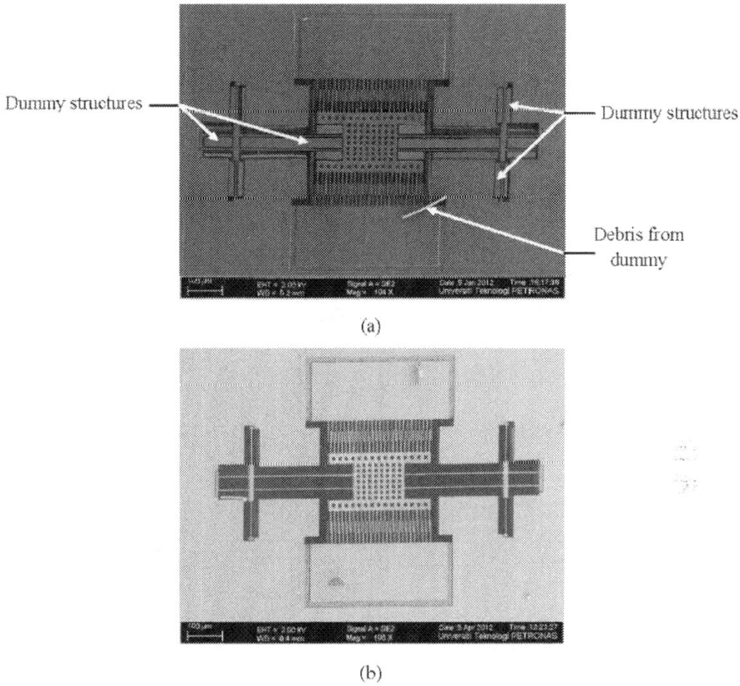

Figure 15. FESEM images of (a) unreleased device showing dummy structure at a large etch holes and (b) fully released structure.

Test structures are used to extract design rules. A large opening undercut rule specifies the maximum undercut that can be achieved by the process. Any microstructure wider than this value will not be released. This rule also specifies the minimum size requirement of the anchor layout, and the skirt for bonding pads. Bonding pads are usually on the periphery of the chip for easy access during bonding and no crossing of bonding wires on the top of the chip. Therefore, at least one side of bonding pads faces large opening and is lateral undercut during the release. The sites that experience forces during wire bonding should avoid these undercut regions and bonding pads need extra area to compensate this lateral undercut. Moreover, this large opening rules specifies the minimum space required from the edge of the etch pit to preserve Si area containing circuitry during micromachining. One way to obtain this data is to use FESEM to measure the undercut from the periphery of the chip. Another way is to use an optical microscope to check which size of square pattern has been released as illustrated in Fig. 16 (a).

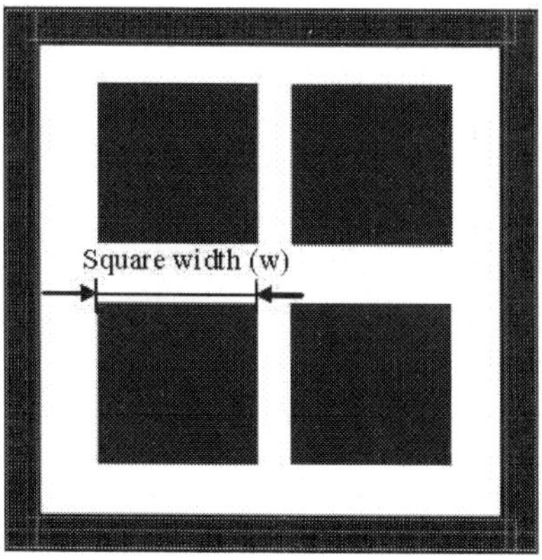

(a) The square structure illustrating the large opening undercut rule

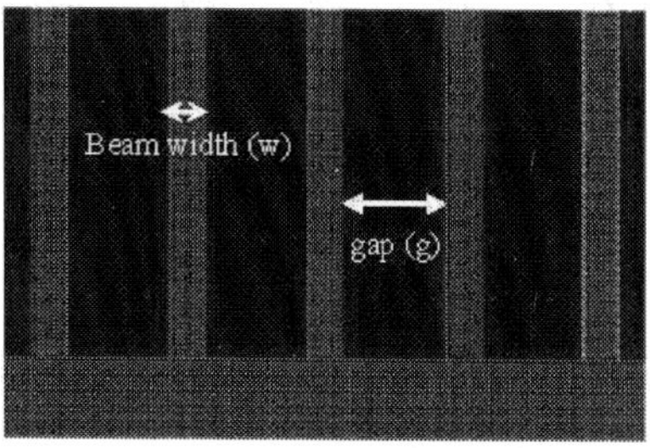

(b) The cantilever structure illustrating the gap undercut rule

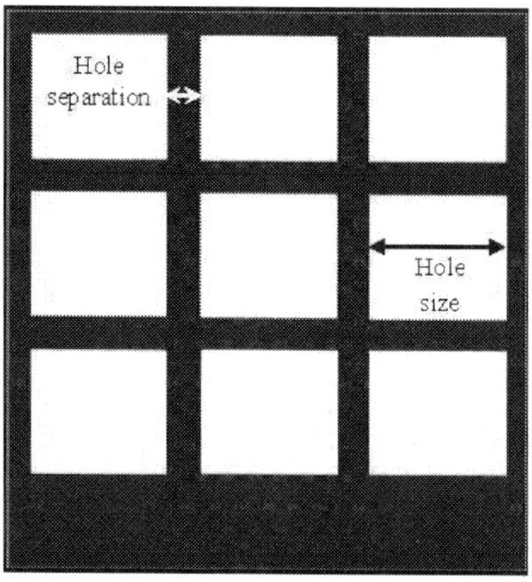

(c) The plate structure illustrating the hole undercut rule

Figure 16. Layout of three test structures used for extracting MEMS design rules for post-CMOS micromachining.

The undercut value is half the size of square width. A gap undercut rule is required for releasing MEMS structures such as a comb finger or a spring. The test structure for this rule is a set of cantilevers with the undercut mainly determined from the sides of the beams. To follow the situation in real layout, such as comb fingers, multiple cantilever beams with equal space between them are used as shown in Fig. 16 (b). The beam width (w) represents the width of the structure to be released. The gap (g) represents the separation between the structures. The beam should be long enough that the undercut from the tip of the cantilever doesn't dominate the etch. Since the residual stress induces out-of-plane curling upon release, the release can be easily detected by an interferometric measurement. A hole undercut rule is required to release plate structures. The test structure for this design rule is a set of plates with release holes. These plates are anchored by four thin beams as shown in Fig. 16 (c). Such plates should be large enough to ensure that the release of the plate is determined by the etch release holes and not from the periphery of the plate.

Similar to the cantilever test structures, the curling after the release of plates can be detected by the interferometric measurement. Extra processing time requires layouts to be modified such that the circuitry is far away from the edge of the mask, all the anchors are wide enough, and all bond pads have enough margins for bonding and probing. The spacing between the squares and the spacing from a square to the surrounding anchor mask area are the size of the square. This is to ensure that squares are uniformly etched from all directions. For each group of cantilever test structures, as in Fig. 16 (b), the widths of all cantilever beams are the same value. The spacing between beams and from tips of cantilever beams to anchor mask areas are the same value [14].

DRIE BASED CMOS-MEMS DEVICES

In this section, two devices fabricated using CMOS-MEMS bulk micromachining technology are described. The devices are a resonant MEMS magnetic field sensor and a resonant MEMS chemical sensor.

Resonant MEMS Magnetic Field Sensor

The post-CMOS process steps of the resonant magnetic field sensor using DRIE were designed to successfully release the sensor structure. The first step in the post-CMOS process was at wafer level with the selective application of photoresist at the back-side of the 8" inch silicon wafer to etch from the back-side to produce a substrate thickness of approximately 50 μm as shown in Fig.17. DRIE was performed using SS110A Tegal plasma etcher to anisotropically etch the silicon substrate to the desired thickness.

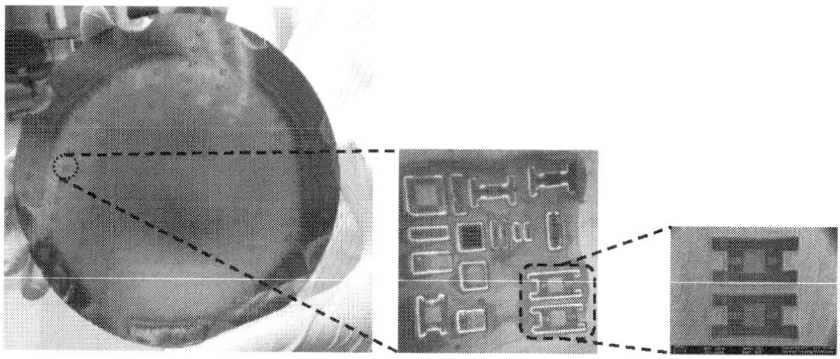

(a) (b) (c)

Figure 17. (a) Optical image of back side Si etching at wafer level, (b) Backside optical image of a chip and (c) FESEM of backside of a sensor.

Optical microscope was used to estimate the thickness of the Si substrate during the etching process.Fig.17 (c) shows the back-side view of one of the sensors that has been successfully etched to approximately 284 μm depth leaving a about 50 μm thickness under the CMOS layers. From the back-side etching of the bulk silicon, the silicon etching rate was found to be ~3.66 μm/min. The second step was dicing the wafer as shown in Fig. 18 (a). Dicing of the wafer has to be done before the front side RIE of SiO_2 and DRIE of silicon is implemented. This sequence was followed in order to prevent the breakage of the released structures.

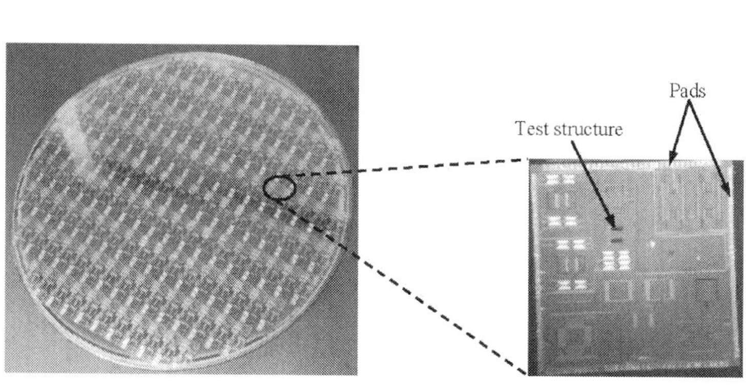

(a) (b)

Figure 18. (a) Optical image of the diced wafer and (b) optical image of front side of a 5 mm × 5 mm chip.

The third process was SiO_2 RIE on the diced chip shown in Fig. 18 (b), which was performed from the front-side of the chip using Tegal plasma etcher. Front side RIE process opened the pattern of the resonant magnetic sensor by removing the SiO_2 layer and thus exposing the silicon underneath while the MEMS sensor pattern was maintained by metal 3 acting as a mask for the pattern. The process at chip level started by flipping over the chips such that the thin film was now on the front side. The chips were then attached to the 8" platinum coated carrier wafer using kapton tape as illustrated in Fig.19. After 13 minutes of etching, a gray colour (silicon colour) appeared in the trenches that indicated the completion of SiO_2 etching. The test structures to the right of the sensor beam were used to estimate SiO_2 etching rate and to avoid over-etch of the SiO_2 material as shown in Fig.18 (b). The total SiO_2etching depth was approximately 5 μm.

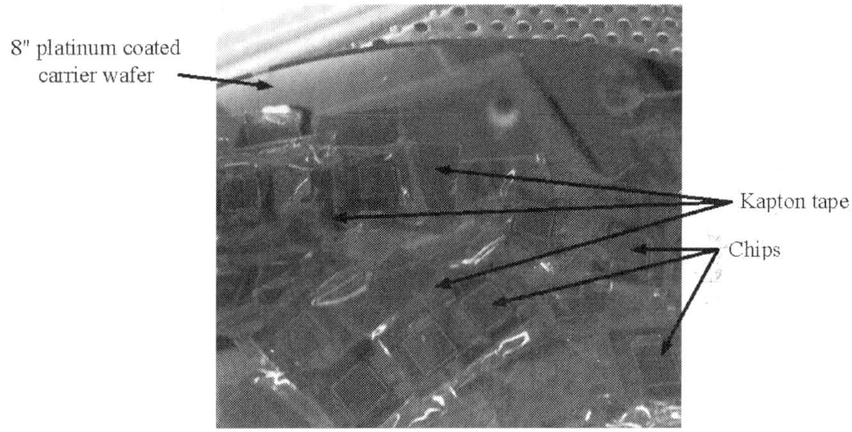

Figure 19. Sample preparation for front side etching.

Next, front-side bulk silicon DRIE process step was performed using the etching rate of 3.66 μm/min obtained from previous backside DRIE etching of silicon. The thickness of the central shuttle and long beams was approximately 50 μm, which was observed under optical microscope during back-side etching. Therefore the test structures on the chip were used as reference. Fig.20 shows FESEM micrograph of the fabricated sensor with inset showing a close-up of the comb fingers and part of dummy structure close to the beams that has fallen out while others still remain intact on the substrate [10]. Additional 3 more minutes were required to fully etch-through. This happened when all the dummy structures dropped and indicated that the etched-through process was completed.

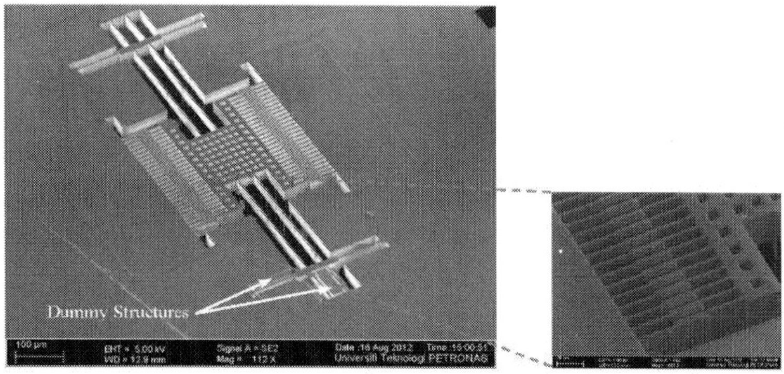

Figure 20. FESEM image of the fabricated CMOS-MEMS resonant magnetic field sensor with inset showing shuttle and stator fingers.

Resonant MEMS Chemical Sensor

In resonant MEMS chemical sensors, known as gravimetric sensors, the principle of detection of the gaseous species is based on the change in resonant frequency of the microresonator membrane or plate. This frequency change results from a change in the mass of the microresonator due to absorption/adsorption of an analyte molecule onto the surface of the active material deposited on it. Fig. 21 shows 2-D schematic diagram of the chemical microsensor device.

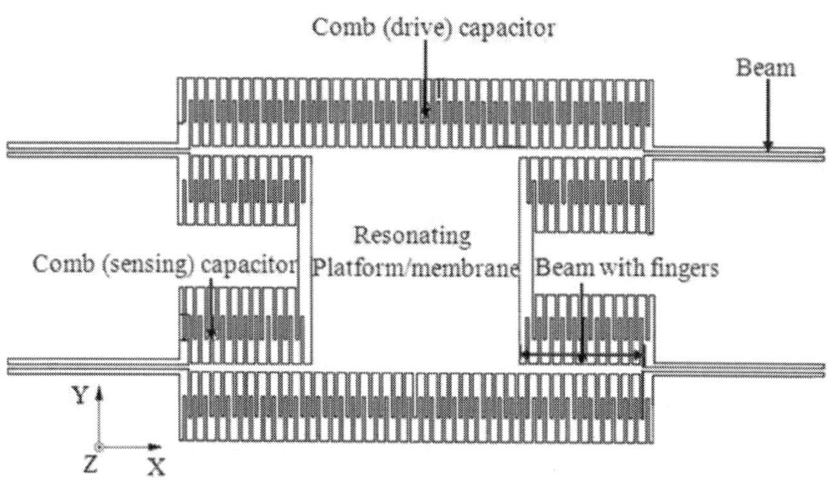

Figure 21. D Schematic of the resonant MEMS chemical sensor.

The post-CMOS micromachining process steps of the resonant MEMS chemical sensor were similar to the process steps of the resonant MEMS magnetic field sensor. The process started with the selective application of photoresist at the back-side of the die around the sensor followed by sample placement on the platinum coated carrier wafer. The second step of the process was to perform back-side silicon DRIE to achieve proof mass thickness of approximately 40 μm. The same plasma silicon etcher was used to anisotropically etch the silicon substrate to the desired thickness. Fig. 22 shows the FESEM image of the successfully released MEMS chemical sensor with the inset showing a close-up view of the perfectly flat sensing comb fingers with the SCS underneath [15].

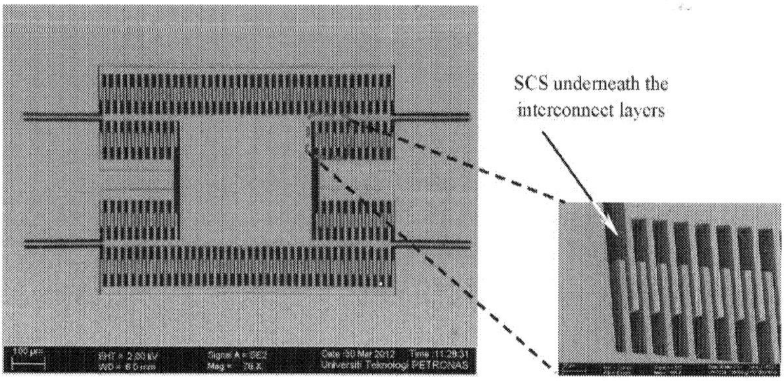

Figure 22. FESEM image of fabricated device with inset showing a close-up view of the fingers with the SCS underneath.

CONCLUSION

This chapter discussed bulk micromachining technology with particular emphases on DRIE post CMOS MEMS bulk micromachining. The chapter was divided into three sections. In the first section an introduction to bulk micromachining of silicon and isotropic and anisotropic wet and dry etching was given. The second section discussed briefly DRIE post-CMOS micromachining process with particular emphasis on DRIE post-CMOS bulk micromachining process and the third and last section provides a few examples of devices fabricated by our research group using the DRIE CMOS-MEMS process. These devices were resonant MEMS magnetic field sensor and resonant MEMS chemical sensor. The aim of the chapter was to discuss and analyze practical processes involved in the design of micromechanical devices using 0.35 μm CMOS technology.

ACKNOWLEDGEMENTS

The authors would like to thank MIMOS Bhd Malaysia for facilitating the microfabrication of the sensors and MOSTI Malaysia for financially supporting this research under E-Science project No.04-02-02-SF0095.

REFERENCE

1. Hongwei Qu and Huikai XieProcess Development for CMOS-MEMS Sensors with Robust Isolated Bulk Silicon Microstructures", *IEEE/ASME Journal of Micro-Electro-Mechanical Systems*, 1611521161
2. http://www.memsnet.org/about/fabrication.html
3. http://www.meengr.ccny.cuny.edu/bridge/material/joined%20files.pdf
4. http://www.mems.ac.cn/jishuwenxian/2009Etching_Processes.html
5. H Xie, L Erdmann, X Zhu, K. J Gabriel, and G. K Fedder, Post-CMOSprocessing for high-aspect-ratio integrated silicon microstructures," Journal of Microelectromechanical Systems, 11200293101
6. G. K Fedder, S Santhanam, M. L Reed, S. C Eagle, D. F Guillou, M. S. C Lu, and L. R Carley, Laminated high-aspect-ratio microstructures in a conventional CMOS process," The Ninth Annual International Workshop on Micro Electro Mechanical Systems, San Diego, CA, 19961318
7. F Ahmad, J. O Dennis, M. H. M Khir, and N. H Hamid, A CMOS MEMS Resonant Magnetic field Sensor with differential Electrostatic actuation and Capacitive sensing," Advanced Materials Research Journal, vols 403-408, 420542092012
8. M Haris, and H Qu, A Cmos-mems, Piezoresistive Accelerometer with Large Proof Mass", IEEE NEMS 2010Xiamen, China, Jan 2010.
9. H Xie, Gyroscope and micromirror design using vertical-axis CMOS-MEMSsensing and actuation," Ph. D. Thesis, Department of Electrical and Computer Engineering, Carnegie Mellon University, Pittsburg, PA, 2002
10. F Ahmad, J. O Dennis, M. H. M Khir, and N. H Hamid, Lorentz force based CMOS Resonant Piezoresistive Magnetic Sensor with differential electrostatic actuation" MY. Patent, Jan.16, 2012
11. Farooq AhmadDesign and simulation of mechanical behavior of MEMS-based resonant magnetic field sensor with piezoresistive output", 2010International Conference on Mechanical and Electrical Technology, 09/2010

12. G. K Fedder, Tri-axial high-g CMOS-MEMS capacitive accelerometer array", 2008IEEE 21st International Conference on Micro Electro Mechanical Systems, 01/2008.
13. X Zhu, Post-CMOS Micromachining of Surface and Bulk Structures," Ph. D. Thesis, Department of Electrical and Computer Engineering, Carnegie Mellon University, Pittsburg, PA, 2002
14. Brand, CMOS-Based Resonant Sensors", IEEE Sensors 2005
15. Y Ahmed, J. O Dennis, M. H. M Khir, and M. N. M Saad, A Method, Apparatus and Sensor for Reversible Detection of Analyte Molecule" MY. Patent, Dec.16, 2011

CITATION

John Ojur Dennis, Farooq Ahmad and M. Haris Khir (2013). CMOS Compatible Bulk Micromachining, Advances in Micro/Nano Electromechanical Systems and Fabrication Technologies, Assistant Professor Kenichi Takahata (Ed.), ISBN: 978-953-51-1085-9, InTech, DOI: 10.5772/55526.

CHAPTER 5

Magnetic Mn-Doped Ge Nanostructures

Faxian Xiu

Department of Electrical and Computer Engineering, Iowa State University, Ames, IA 50011, USA

ABSTRACT

With the seemly limit of scaling on CMOS microelectronics fast approaching, spintronics has received enormous attention as it promises next-generation nanometric magnetoelectronic devices; particularly, the electric field control of ferromagnetic transition in dilute magnetic semiconductor (DMS) systems offers the magnetoelectronic devices a potential for low power consumption and low variability. Special attention has been given to technologically important group IV semiconductor based DMSs, with a prominent position for Mn doped Ge. In this paper, we will first review the current theoretical understanding on the ferromagnetism in Mn_xGe_{1-x} DMS, pointing out the possible physics models underlying the complicated ferromagnetic behavior of Mn_xGe_{1-x}. Then we carry out detailed analysis of Mn_xGe_{1-x} thin films and nanostructures grown by molecular beam epitaxy. We show that with zero and one dimension quantum structures, superior magnetic properties of Mn_xGe_{1-x} compared with bulk films can be obtained. More importantly, with Mn_xGe_{1-x} nanostructures, such as quantum dots, we demonstrate a field controlled ferromagnetism up to 100 K. Finally we provide a prospective of the future development of ferromagnetic field effect transistors and magnetic tunneling junctions/memories using dilute and metallic Mn_xGe_{1-x} dots, respectively. We also point out the bottleneck problems in these fields and rendering possible solutions to realize practical spintronic devices.

INTRODUCTION

Dilute magnetic semiconductors (DMSs) attract tremendous interest as emerging candidates for the microelectronics industry due to their uniqueness in exhibiting spin-dependent magneto-electro-optical properties. Their distinctive material characteristics such as spin-dependent coupling between semiconductor bands and the localized states promises magnetoelectric effect—the modulation of magnetic properties by an applied electric field—in semiconductors [1, 2]. Thus, a wide variety of semiconductor devices can be envisaged [3], such as spin polarized light-emitting diodes, lasers [4, 5] and spin transistor logic devices [2, 6,7]. The development of these devices is considered as a possible route for extending the semiconductor scaling roadmap to spin-added electronics [8].

DMS materials can be developed by alloying semiconductors with several percentages of magnetic transition elements (with 3-d orbitals), such as Fe, Mn, Co, Ni, and V [9]. The anticipated outcome due to this doping is the fact that the transition elements occupying substitutional sites hybridize with the semiconductor host via the sp-d exchange interaction and help to enhance the spin-dependent transport, which in turn increase both the magnetization and the Curie temperature (Tc) of the semiconductor system [10]. This means that the transition metal doping can generate strong spin-dependent coupling states in semiconductor systems, which may be further modulated by an electric field [2]. Since the band gap engineering and crystal structures of a DMS material are compatible with other semiconductors, they offer significant integration advantages as well [3].

The dependence of the transition temperature on the density of mobile carriers is considered as a good evidence for a genuine DMS. However, the key challenge for the growth of DMS materials is the ability to control synthesis of transition metals in semiconductors, aimed at increasing the doping concentration in order to realize room-temperature ferromagnetism. Thus, achieving both semiconducting and ferromagnetic property at room temperature has become a great challenge. Unfortunately, this is hampered by low solubility of transition metal species with a strong tendency to form metallic clusters or heterogeneous regions within the matrix. So the primary research on DMS growth is to produce samples with uniform doping and being free of metallic precipitates.

Substantial theoretical and experimental work were carried out in the past two decades, either to predict or to measure the magnetic properties on a variety of semiconductor materials, including binary and ternary compounds such as III-V, II-VI, IV-VI, VI oxides, III nitrides, and group IV systems. Among them, (Ga, Mn)As became the first and the mostly studied DMSs since 1996 [11]; but to our knowledge, some studies lacked the threshold doping profile to push its curie temperature to room temperature and above (currently around 190 K [12–15]). Since then, substantial research has been carried out in III-V and II-VI semiconductors to improve the Tc by increasing the solubility limit and meanwhile minimizing self-compensation effect of magnetic structures, using radical techniques such as delta doping of modulation-doped quantum structures [16–19]. Another approach to achieve high-Tc DMS is to use transition metal-doped wide bandgap nitrides and oxides, such as GaN, AlN, ZnO, and TiO_2. In these materials, high magnetic impurity concentrations are needed due to the lack of long-range interaction among magnetic spins. However, there have been only limited reports on the carrier mediated effect in the wide band gap materials at room temperature [20–22]. Though considerable amount of work has been done on III-V and II-VI DMSs [2, 23], the lack of high Tc and the difficulty in achieving electric field modulated ferromagnetism (carrier-mediated ferromagnetism) at or above room temperature deters them from being considered for semiconductor integration.

Group IV semiconductors are of particular interest to the spintronics technology because of their enhanced spin lifetime and coherent length due to their low spin-orbit coupling and lattice inversion symmetry [24, 25]. There is also considerable interest for transition metal-doped group IV semiconductors [26–36] to the semiconductor industry, owing in part to their excellent compatibility with silicon process technology. Several transition metals—including the use of single and codoping of elements such as Mn, Cr, Co, Fe, Co-Mn, and Fe-Mn—have been used as magnetic dopants in both Si and Ge either through blanket implantation or through in situ doping during epitaxial growth process. Both these materials are reported to be ferromagnetic [33, 37, 38]. The reported experimental results also clearly showed the presence of carrier-mediated ferromagnetism in DMS Ge [39, 40]. A recent study predicted that the Tc could be increased by enhancing the substitutional doping of Mn in Ge and Si, via codoping—by adding conventional electronic dopants such as As or P during the Mn doping process [41]. Theories [42] and experiments [29, 43] further suggest that the ferromagnetic transition temperature is related to the ratio of interstitial to substitutional Mn similar to that in the group III-V system [10].

Considering the fact that, nanosystems such as nanowires and quantum dots (QDs) are the versatile building blocks for both fundamental studies in nanoscale and the assembly of present day functional devices, low-dimensional Ge DMS systems such as Mn_xGe_{1-x} nanowires [13, 30, 44, 45] and QDs [46] have also been reported to have room temperature ferromagnetism. The above results are supported by modeling results [47] although carrier-mediated exchange has not yet to be experimentally verified in nanowires. Similarly, high Curie temperatures have been observed in other transition metal-doped semiconductor nanostructures such as II-VI (including ZnO [48–50], ZnS [51, 52], CdS [52]), and III-V (including GaN [52, 53], GaAs [54]), but only inconclusive reports have been published on the magnetoelectric transport in these nanostructures. It could be attributed to the complicated ferromagnetic properties caused by precipitates, second-phase alloys and nanometer-scale clusters. Therefore, it is important and indispensable to differentiate their contributions. Unfortunately, there is still a lack of fundamental understanding of the growth process and spin-dependent states in these systems that gives an impetus to extensive investigations into these and related material systems.

In this paper, we will first introduce the state-of-art theoretical understanding of ferromagnetism in group IV DMSs, particularly pointing out the possible physics models underlying the complicated ferromagnetic behavior of MnGe. Then, we primarily focus on the magnetic characterizations of epitaxially grown Mn_xGe_{1-x}thin films ranging from three-dimensional thin films to zero-dimensional QDs. It is found that when the material dimension decreases (to zero), it evidences a change of material structures being free of precipitates and a significant increase of Tc over 400 K. The preliminary explanation on these behaviors can be attributed to the carrier confinement in the DMS nanostructures which strengthens hole localization and subsequently enhances the thermal stability of magnetic polarons, thus giving rise to a higher Tc than those of bulk films.

Another important aspect of DMS research lies in the electric-field-controlled ferromagnetism. We emphasize the existing experimental efforts and achievement in this direction and provide a detailed description on the field-controlled ferromagnetism of the Mn_xGe_{1-x} nanostructures. We show that the ferromagnetism for the Mn_xGe_{1-x} nanostructures, such as QDs, can be manipulated via the control of gate voltage in a gate metal-oxide-semiconductor (MOS) structure (up to 100 K). These experimental data suggest a hole-mediated ferromagnetism as controlled by a gate bias and promise a potential of using Mn_xGe_{1-x} nanostructures to achieve spintronic devices. Beyond the scope of DMSs, we will present the metallic Mn_xGe_{1-x} with x ~ 20%, in which a constant Curie temperature was observed. The emphasis, however, is the manipulation of these metallic dots into self-assembled periodic arrays for potential device applications. Finally, we render general comments on the development of the Mn_xGe_{1-x} system and possible research schemes to develop spintronics devices based on this material system.

THEORIES OF FERROMAGNETISM IN GROUP IV DMS

As in other popular dilute magnetic semiconductors such as (Ga, Mn)As, the formation of magnetic order and the origin of ferromagnetism in Mn_xGe_{1-x} can be described (within the parametric limits) either using complementary mean field approximation models such as the Zener-Kinetic exchange [55], the Ruderman-Kittel-Kasuya-Yoshida (RKKY) [56] interaction, and percolation theory [57] or using density functional theory such as ab initio full-potential augmented plane wave (FLAPW) [58] electronic calculations. Interestingly, the Zener-kinetic exchange model proposed in 1950 could be used for interpreting many of the experimental results in transition metal-doped semiconductors [59–61] besides the transition metals themselves. All the above theoretical approaches have tried to address and to a certain extent successfully answered chemical, electronic, and magnetic properties of the transition metal-doped group IV semiconductors within the context of (1) the nature of transition metal impurities, (2) the origin of ferromagnetism, (3) the influence of dopants on the Tc, (4) the solubility limits of carrier density on the chemical nature, (5) the influence of carrier density on the electronic structure, and (6) the influence of different hosts on the electronic structure. Within the above theoretical approaches, these models assume that the ferromagnetism in transition metal-doped Ge system is mediated through exchange coupling between the carriers and the interacting lattice. This means that, the interaction is through p-d orbital hybridization between the d levels and the valence band p states of Ge. According to these models, as the number of dopant carriers increases both the electronic mobility and the ferromagnetic ordering of the system also increases along with the Tc. Thus, the assumption is that the strong exchange coupling between the p and d orbitals leads to strong kinetic exchange coupling between the spin-polarized holes and spin polarized TM substitutional vacancies. Among all the 3-d TM impurities, [TM = V, Cr, Mn, Fe, Co, Ni], Mn is the most favored dopant. This is because compared to other dopants, Mn favors less clustering, low nonuniform doping distribution and higher substitutional doping over interstitials. The ab initio electronic structure calculations within density functional theory show that such doping behaviors could result in relatively higher carrier hole concentrations and increased magnetic moments [27, 58].

Complementary theoretical work using the frozen-magnon scheme also shows that the ferromagnetism is produced only through holes and Mn is a good source for generating holes in Ge system [62].

In addition, some of the other factors that determine the ferromagnetic strengths in Ge are (1) disorder of Mn site locations, (2) distance between Mn-Mn atoms, (3) solid solubility limits and preferential surface orientation for interstitial and substitution site formation [63–65], and (4) codoping of Co or Cr with Mn. (They were found to reduce cluster formation but not to significantly contribute to the ferromagnetism in the system [66, 67].) According to full potential linearized augmented plane wave (FLAPW) calculations made on the RKKY model, the coupling between the Mn atoms could be ferromagnetic (FM) or antiferromagnetic (AFM) in nature but the interaction is very localized to Mn sites and depends on the distance between Mn atoms: the density of states decreases rapidly as the Ge atom moves away from Mn [58]. However, disorder of Mn site locations is found to influence the ferromagnetic strength and the transition temperature. In addition, the impact of inherent disorder is also confirmed both by experiments and by the mean field approximation [57, 68, 69]. Some of the above models found themselves hard to explain the long-range exchange interaction and coupling strength when there is varying dopant concentration and disorder ranging from high resistive state to half metallic state. On the other hand, there are attempts to find solutions for disorder-dependent magnetic variation using Percolation theory-based mean field approximation calculation [57, 70]. Percolation theory provides possible explanations for the experimentally observed low values of saturation magnetization in low carrier density Mn_xGe_{1-x} because the infinite cluster of percolating bound magnetic polarons triggering the long-range ferromagnetic order leaves out a large number of Mn moments at temperatures less than Tc. However, the role of disorder with respect to insulating and conductive metallic phases and their effects on mobility is still not well understood. Conversely, the FLAPW calculations using the density functional theory shows that Mn favors long-range ferromagnetic alignment over short-range antiferromagnetic alignment and this increases with the increase of Mn content which is in agreement with many of the experimental results [58]. The estimated Curie temperature is within the range of 134–400 K and is in agreement with experimental and theoretical results [47].

However, there are also conflicting reports that long-range FM order could only happen at low temperature (12 K) and that at low temperature MnGe could show only spin glass behavior due to the intercluster formation between the FM Mn-rich clusters [42, 71]. With these reports, it is understood that the magnetic phases are more favored over paramagnetic phases by an energy difference of 200 meV/Mn atom [58]. The report also states that the Mn_xGe_{1-x} system is very close to half metallicity irrespective of the amount of doping concentration [58, 72]. Electronic band structure calculations using density of states (DOSs) show that the binding energy of the Ge 4s states are basically unaffected by the exchange splitting of Mn d states and the valence band maximum of the minority spin channel in Mn_xGe_{1-x} reaches EF at Gamma (T) such that the spin gap becomes zero [58, 72]. In other words, the DOS shows a valley around EF and is strictly zero at EF in the minority spin channel. On the other hand, the energy gap is indirect and the bands around EF in the majority spin channel arise from Mn d and Ge p hybridizations so as to give rise to hole pockets closer to the gamma point with a localized magnetic moment of about 3 μ_B. This could also be explained using chemical and structural configurations of Mn in Ge. The different chemical species determine the relative positions of the anion and cation atomic energy levels in the Mn_xGe_{1-x} while the different anion size dictates essentially the equilibrium lattice constants [58]. This essentially explains the carrier-mediated effect in Mn_xGe_{1-x} even though the system is closer to half metallic. The aim of all these theoretical works is to build foundation for achieving carrier-mediated ferromagnetism at temperatures above room temperature. It is understood that this can be achieved only by growing cluster-free and uniformly doped Mn_xGe_{1-x} samples. Experimental reports show that Tc as high as 400 K is obtained for cluster-free Mn_xGe_{1-x} samples grown (1) using subsurfactant epitaxial method [73], (2) epitaxial QDs [40], and (3) nanowire structures [13, 30, 44]. However, there are only very few theoretical studies depicting the relation between the quantum confinement and its influence on the origin of ferromagnetism in low-dimensional DMS structures [74–76]. It apparently lacks of systematic theoretical study particularly on Mn_xGe_{1-x} QDs. Therefore, it will be interesting to understand the ability of the quantum confinement phenomenon to retain spin polarization in Mn_xGe_{1-x} QDs structure. Studies on III-V QDs using LSDA approximation show that, unlike in the bulk structures, adding a single or multiple carriers in a magnetic QD can both strongly change the total carrier spin and the temperature of the onset of magnetization and this can be modulated by modifying the quantum confinement and the strength of coulomb interactions. These theoretical results must have ramification in the high Curie temperature Mn_xGe_{1-x} QDs as well since our studies also reveal similar characteristics. But, all of the effects including high Tc that we observe cannot be explained without invoking tiny defects either at the interface or on the surface although the conventional transmission electron microscopy (TEM) cannot reveal their existence due to the limited resolution. All of the above material and electronic properties have been studied in order to understand how feasible a DMS material is for carrier-mediated ferromagnetism.

GROWTH AND CHARACTERIZATIONS OF GROUP IV DMS AND NANOSTRUCTURES

Overview of the Mn-Doped Ge

As mentioned above, the Ge-based DMS has attracted extensive attention due to its possibility to be integrated with the mainstream Si microelectronics, in which they may be used to enhance the functionality of Si-integrated circuits [27]. In particular, the hole-mediated effect discovered in Mn_xGe_{1-x} DMS opens up tremendous possibilities to realize spintronic devices with advantages in reducing power dissipation and increasing new functionalities, leading to perhaps normally off computers. To date, there are many reports on the Mn_xGe_{1-x} growth and characterizations by molecular beam epitaxy (MBE) [27, 28, 31, 33, 43, 77–85], ion implantation [86–93], and bulk crystal growth [37, 94]. We can divide them here, for an easier orientation, into two groups: those that cover fundamental studies of phase formation, ferromagnetism, and transport properties [26, 27, 29, 31, 33, 35–37, 43–45, 64, 77–85, 94–115], and those that aim to minimize cluster formation and enhance Curie temperatures via codoping methods [28, 32, 34, 116–118].

Experimental results show that the Mn-doping process in Mn_xGe_{1-x} is complex. Both the Tc and the saturation magnetization depend on the interplay of a variety of factors, which are ultimately determined by growth conditions and postannealing process [33]. The concentration and distribution of Mn dopants, the carrier density, the presence of common defects such as Mn interstitials and Mn clusters significantly influence the magnitude and interactions of the magnetic coupling [94]. In early 2002, Park et al. [27] demonstrated ferromagnetic Mn_xGe_{1-x} thin films by low-temperature MBE. It was found that the Mn_xGe_{1-x} films were p-type semiconductors in character with hole concentration of 10^{19}~10^{20} cm^{-3} and exhibited pronounced extraordinary Hall effect. The Tc of the Mn_xGe_{1-x} thin films increased linearly with Mn concentration from 25 to 116 K (Mn $\leq$ 3.5%). Field-controlled ferromagnetism was also observed in a simple gated structure through application of a gate voltage ($\pm$0.5 V), showing a clear hole-mediated effect at 50 K. Such a small gate voltage presents an excellent comparability with conventional low voltage circuitry. The origin of ferromagnetic order was understood in the frame of local-spin density approximation (LSDA), where strong hybridization between Mn *d* states of T_2 symmetry with Ge *p* states leads to the configuration e(↑)2T2(↑)2T2(↓)1, with a magnetic moment of 3 μ_B. However, the LSDA calculations gave overestimated *Tc* which was attributed to the incomplete activation of Mn in experiments and the absence of hole compensation in the simulation [27].

Since then, various preparation techniques were employed in order to produce Mn-doped Ge DMS, including aforementioned MBE [28, 31, 33, 43, 77–85], single-crystal growth [37, 94], and ion implantation [86–93], aiming to further increase *Tc* and to obtain the electric field controllability at room temperature [94]. Among these efforts, Cho et al. [37] reported the synthesis of Mn-doped bulk Ge single crystals with 6% of Mn and a high ferromagnetic order at about 285 K via a vertical gradient solidification method. The origin of the high *Tc*was found to be complex because of the presence of dilute and dense Mn-doped regions. Jaeger et al. [77], however, pointed out that the magnetic properties of these samples were clearly dominated by the presence of the intermetallic compound $Mn_{11}Ge_8$. This was indeed observed by Biegger et al. [94], where similar Mn_xGe_{1-x} crystals were produced via Bridgman's crystal growth technique and intermetallic compounds were found in both Mn-rich and Mn-poor regions. These experiments suggest that the bulk Mn_xGe_{1-x} crystals produced at over 1000°C may not be an appropriate method for the Mn_xGe_{1-x} DMS preparation [94].

Apart from the bulk crystal growth, MBE has been widely recognized as a power tool for DMS growth since it can provide nonequilibrium conditions to enhance the incorporation of Mn dopant [27]. Via this technique, Mn doping was found to be extremely sensitive to growth conditions, particularly to the growth temperature. Excellent work on this aspect can be found in references [32, 33, 43, 79, 81, 83, 85]. In general, at low growth temperatures (Tb<120∘C), the diffusion of Mn atoms leads to the formation of Mn-rich nanostructures, such as nanocolumns [82] and nanodots [81] with irregular shapes. These nanostructures could contain a high Mn concentration up to 38% 33, while the surrounding matrix remains a low Mn concentration less than 1%. A perfect lattice coherence was found in the proximity of the nanostructures, indicative of a strong compression for coherent nanostructure and a large tension for the surrounding matrix [82]. These observations are in good agreement with theoretical predictions of a spinodal decomposition [119], which occurs under a layer-by-layer growth mode with strong pairing attraction between Mn atoms and a tendency of surface diffusion of Mn atoms. These theoretical calculations successfully explained the formation of nanostructures in (Ga, Mn)N and (Zn, Cr)Te) [82, 119], which can be also readily applied to the Mn_xGe_{1-x} system [82]. The magnetic properties, however, exhibit various characteristics depending on the nanostructure size and the Mn concentration. In most cases, zero-field-cooled (ZFC) and field-cooled (FC) magnetizations show superferromagnetic properties with a blocking temperature of ~15 K. It is now clear that such a low blocking temperature is originated from the coherent nanostructures with a diameter below several nanometers. Curie temperatures of these nanostructures were also identified to be below 170 K and varied with different Mn concentration. No Mn_5Ge_3 metallic phases were observed at low temperature regime (Tb<120∘C).

Bougeard et al. [81] also found that when the dimension of the nanostructures further reduced, the films showed no overall spontaneous magnetization down to 2 K. TEM results were interpreted in terms of an assembly of superparamagnetic moments developing in the dense distribution of nanometer-sized clusters. In addition to the single crystalline Mn_xGe_{1-x}, homogeneous nanoclusters were also observed under certain growth conditions and contributed to ferromagnetic order below 100 K [82, 107]. To render a general understanding of the Mn_xGe_{1-x} magnetic properties, Jaeger et al. [77] presented a detailed study on $Mn_{0.04}Ge_{0.96}$ and $Mn_{0.20}Ge_{0.80}$ thin films grown at low temperatures and related their magnetic behavior to spin-glass. A frozen magnetic state at low temperatures was observed and attributed to the formation of nanostructures. This is in a good agreement with Devillers et al. [82]. Therefore, it can be concluded that the Mn_xGe_{1-x} magnetic semiconductor is not a conventional ferromagnet since its magnetic properties are significantly complicated by the formation of nanostructures and metallic clusters.

When the growth temperature falls into 120~145°C, the Mn_5Ge_3 nanoclusters start to develop and dominate the magnetic property with a Curie temperature of 296 K [33, 82]. This phase, frequently observed at high-temperature growth, is the most stable (Mn, Ge) alloy. The other stable compound $Mn_{11}Ge_8$ was also observed in nanocrystallites surrounded with pure Ge [83]. Although they are ferromagnetic, the metallic character considerably jeopardizes their potential for spintronic applications. Under a narrow growth window around 130°C, another phase $MnGe_2$ nanocolumns could be developed with a ~33% of Mn and a Tc beyond 400 K [33]. A remarkable feature of these clusters is the large magnetoresistance due to geometrical effects. Further increasing the growth temperature leads to the dominating magnetic behavior from the Mn_5Ge_3 clusters. As the coherent nanostructures grow larger at high temperature growth, higher blocking temperatures (~30 K) were also obtained [82].

To minimize the phase separation, Mn codoping with Co was attempted to stabilize structures at high Mn-doping concentration [28, 34, 116]. It showed that codoping with Co can dramatically reduce phase separation and diffusion of Mn within the Ge lattice while it magnetically complements Mn. The measured strain states indicate the critical role played by substitutional Co with its strong tendency to dimerize with interstitial Mn [34]. The highest Tc achieved so far is about 270 K [28]. Similarly, Gareev et al. [118] reported on the codoping of Fe to achieve a Tc of 209 K. Recent polarized neutron reflectivity measurements provided evidence to show no segregation and lower clustering tendencies for higher Fe doping [117], which further supported codoping approach in Mn_xGe_{1-x}.

Ion implantation was also attempted due to its efficiency of Mn incorporation [92]. However, extensive experiments revealed that Mn-rich precipitates were developed and buried in a crystalline Ge matrix [86–93] in a much similar manner to that of the thin films grown by MBE [32, 43, 81, 83].

Although there has been much progress in producing high-quality Mn_xGe_{1-x} thin films, the formation of uncontrollable metallic clusters yet remains a challenge. High Tc DMS thin films without any precipitations seem to be a critical obstacle for the fabrications of practical spintronic devices functioning at room temperature, not to mention the room-temperature-controlled ferromagnetism. Table 1 provides a brief summary of the current status of Mn_xGe_{1-x} on the growth method, the Tc, and the field controllability of ferromagnetism.

Table 1. A brief summary of the MnGe growth and characterizations.

Mn-doped Ge	Growth method	Mn (%)	T_c (K)	Origin of FM	Field controlled FM	Ref.
Thin films	MBE	0.6~3.5	25~116	DMS	50 K	[27]
Bulk single crystal	Sintering	6.2	150~285	Mn_5Ge_3, $Mn_{11}Ge_8$	No	[37, 94]
Thin films	MBE	<6	~300	Mn_5Ge_3	No	[36, 77, 98]
Thin films (with $MnGe_2$)	MBE	6	>400	$MnGe_2$ clusters	No	[33]
Thin films	Implantation	A few	~300	Mn_5Ge_3, $Mn_{11}Ge_8$	No	[39, 88, 89, 92]

Mn_xGe_{1-x} by Molecular Beam Epitaxy

To gain further insight into the complex growth of this material system, we have performed detailed structural and magnetic characterizations of Mn_xGe_{1-x} grown by MBE with different thickness. Our experiments showed that the Mn-doping behavior and magnetic properties are dramatically different when the material dimension becomes smaller. Bulk Mn_xGe_{1-x} films always tend to form clusters no matter what growth conditions are. When it comes to one and zero dimensions, however, the cluster formation may be possibly limited because the Mn concentration cannot accumulate enough and the strain can be easily accommodated to minimize the metallic clusters [84, 93]. We further show that self-assembled Mn_xGe_{1-x} DMS QDs can be successfully grown on Si substrate with a high Tc in excess of 400 K. Finally, the electrical-field-controlled ferromagnetism was also demonstrated in MOS structures using these QDs as the channel layers, as elaborated in Section 4.

Mn_xGe_{1-x} Thin and Thick Films

Mn-doped Ge films were grown on Si substrates by a Perkin-Elmer solid source MBE. Si substrates were cleaned by $H_2SO_4 : H_2O_2$ (5 : 3) and 10% HF with a final step of HF etching. The native oxide was removed by a 10 min flash at 800°C in the vacuum chamber. After that, a direct growth of 80 nm thick Mn_xGe_{1-x} on Si was carried out at 250°C. A 5 nm Ge capping layer was deposited on the surface in order to prevent $Mn_xGe_{1-x}$from oxidizing. To probe the annealing effect, identical Mn_xGe_{1-x} thin films were grown repeatedly and followed by in situ annealing process at 400°C for 30 min. All as-grown and annealed samples were characterized by high-resolution TEM (HR-TEM, FEI Tecnai F30 operating at 200 keV) and superconducting quantum interference device (SQUID, Quantum Design MPMS XL 5). Cross-section HR-TEM specimens along ⟨110⟩ were prepared by using a tripod technique, followed by a final thinning using a Gatan precision ion polishing system.

Figures 1(a) and 1(b) show cross-sectional TEM images of the as-grown and annealed 80 nm thick $Mn_{0.04}Ge_{0.96}$. Nonuniform and discontinued rough films were observed in both cases. Figures 1(c) and 1(d) are their corresponding Mn elemental maps, respectively. Mn is seen to concentrate on top for both samples. Mn-containing clusters are confirmed near the top of the crystalline layer as indicated by the white arrow in Figure 1(a). A HR-TEM image clearly indicates that a Mn-containing cluster is formed on top of the Ge lattice as shown in Figure 1(e). Using the lattice constant of Ge as the reference, the lattice spacings of the Mn-containing cluster can be determined as 0.62 and 0.25 nm, which are matched with the distances of (100) and (002) atomic planes of the hexagonal

Mn_5Ge_3 phase. Fast Fourier transformation (FFT) map shown in the inset of Figure 1(e) also indicates that the cluster belongs to the hexagonal Mn_5Ge_3 phase. Mn_5Ge_3 clusters may be formed when the Mn concentration is sufficiently high for the nucleation of clusters [84].

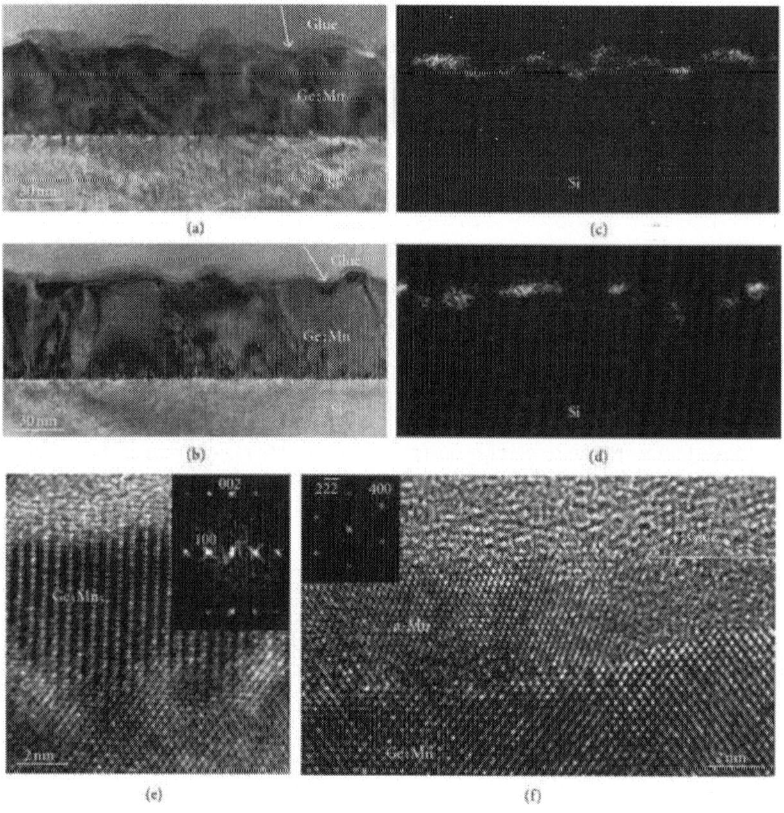

Figure 1. (a) As-grown sample at 250°C, (b) annealed sample at 400°C; (c) and (d) are corresponding Mn maps of (a) and (b), respectively; (e) HRTEM image of (a) typical Mn5Ge3 cluster; (f) HRTEM image of a typical α-Mn metallic cluster epitaxially grown on the (Mn)Ge lattice. The insets in (e) and (f) are the corresponding FFT maps of the images.

Figure 2 shows the TEM cross-section of as-grown and 400°C annealed samples. Stacking faults along {111}atomic planes are clearly observed in both cases. The annealing process does not dramatically affect the stacking-faults. However, the magnetic properties have a noticeable change. Figures 3(a) and 3(d) show clear hysteresis loops at 298 K for both the as grown and annealed samples. Before annealing, the film shows a clear hysteresis at 10 K while a very small magnetic moment is observed at room temperature. After annealing, the hysteresis showed a significant change and the coercive field at 10 K decreased from 430 Oe to 92 Oe. For the as-grown film, Mn_5Ge_3 formed on the surface gave the *Tc* of around 298 K as shown in Figure 3(b). The blocking temperature was estimated to be about 250 K. After annealing, both Mn_5Ge_3 and Mn clusters led to a complex magnetic behavior, showing multiple phase transitions at different temperatures in Figure 3(e). The *Tc* increased from 298 to over 400 K. Further investigation is needed to understand the change of the magnetic property. The Arrott's plots were made to confirm the *Tc* in Figures 3(c) and 3(f). The estimated Curie temperatures for the as-grown and annealed samples were obtained to be 298 K and above 400 K, respectively.

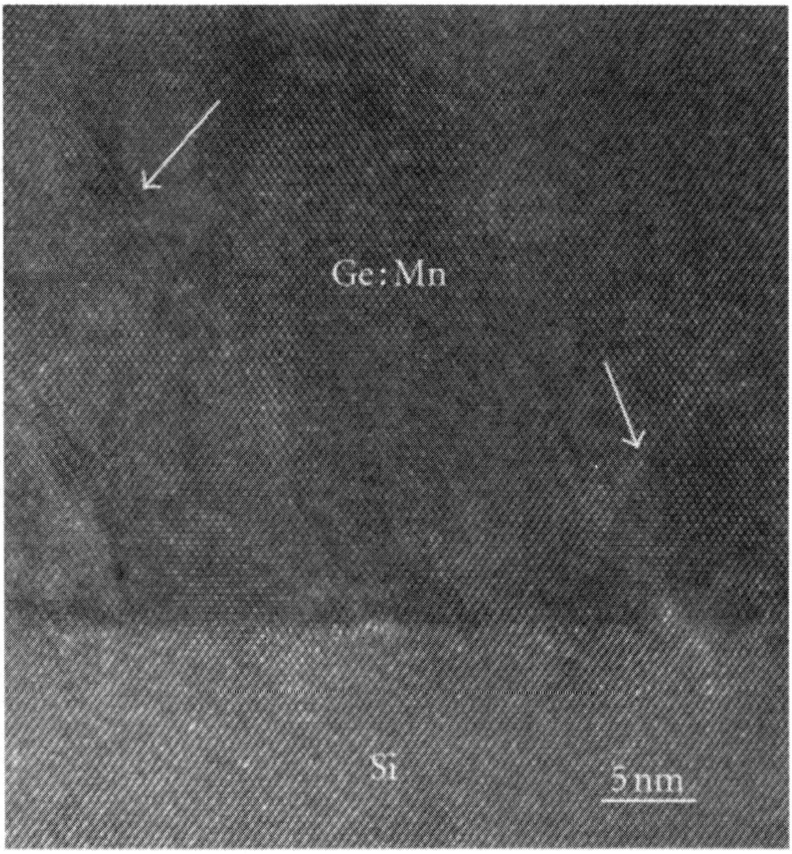

(a)

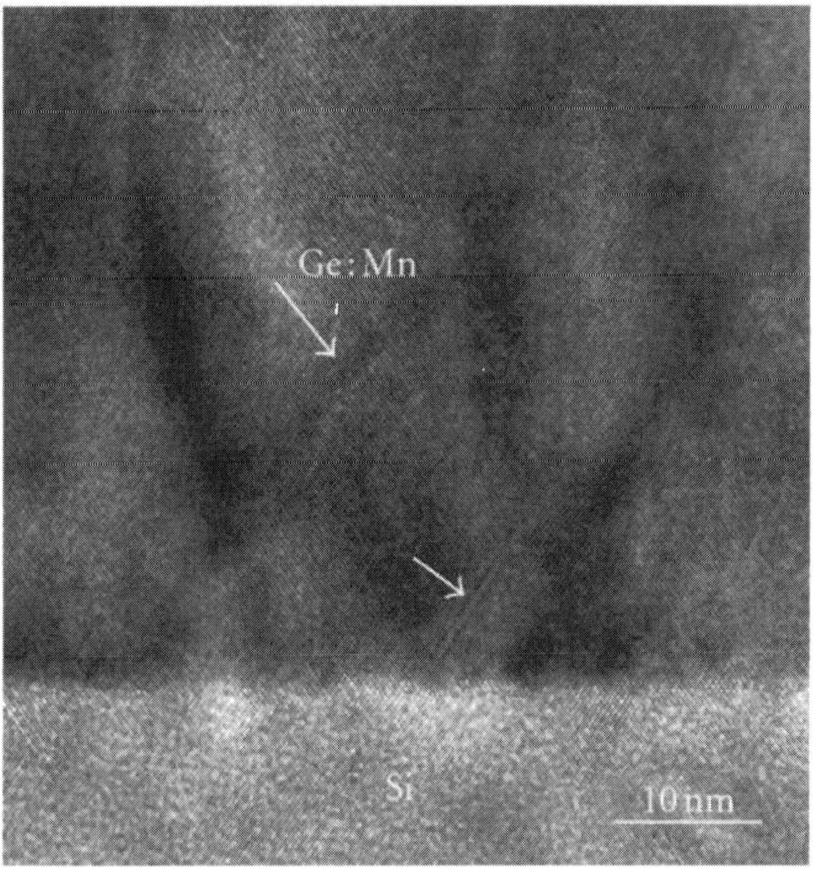

(b)

Figure 2. 80 nm thick Mn-doped Ge grown on Si at 250°C, (a) as-grown film and (b) after 400°C annealing at for 30 min. Stacking faults are marked by arrows.

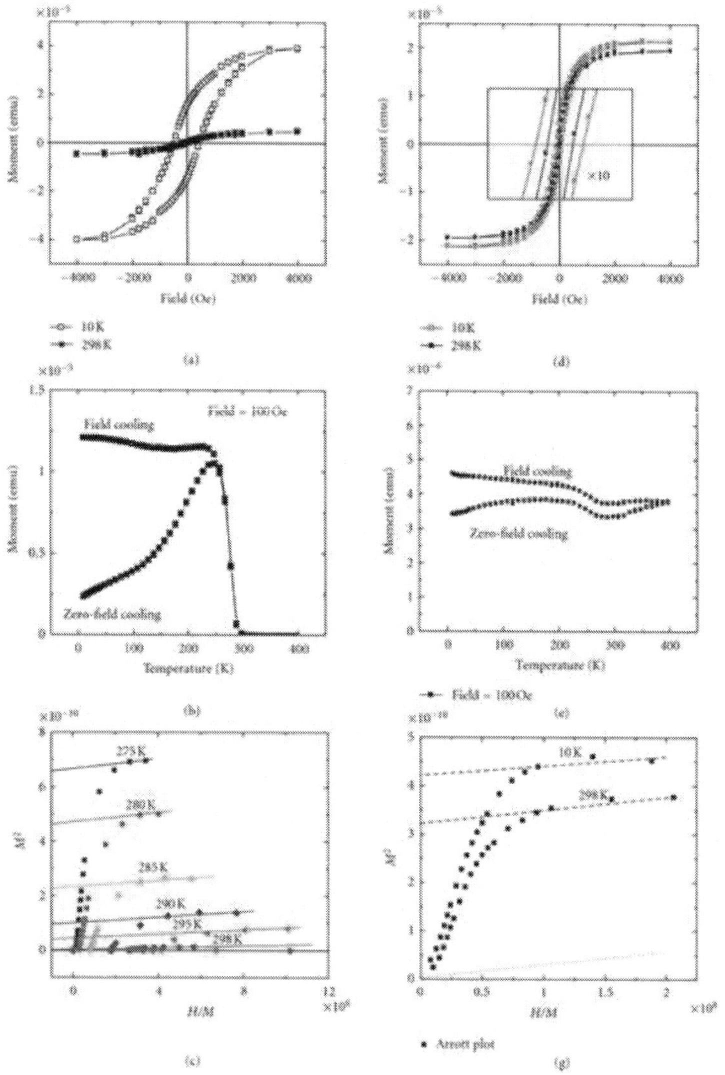

Figure 3. SQUID measurement of 4% Mn-doped Ge grown on Si, the left column is for the as-grown film and the right column is the film after 400°C annealing. (a) and (d) are the hysteresis loops at 10 and 298 K; (b) and (e) are moment versus temperature curves; (c) and (f) are Arrott's plots. The red line in (f) shows the expected Arrott's plot at the Curie temperature.

To understand the magnetic behavior of Mn in an ultrathin Ge, a Mn_xGe_{1-x} film with a thickness of about 15 nm was grown directly on Si substrate by following the same growth procedure as that for the 80 nm thick film. Figures 4(a) and 4(b) show typical cross-sectional TEM images and general morphologies of the as-grown and annealed samples. From these figures, a relative uniform crystalline layer, adjacent to the Si substrates topped with an amorphous layer, can be seen in both cases. The formation of amorphous layers is not clear yet, but it could be attributed to the growth conditions and the sample preparation for TEM when exposing to the air.

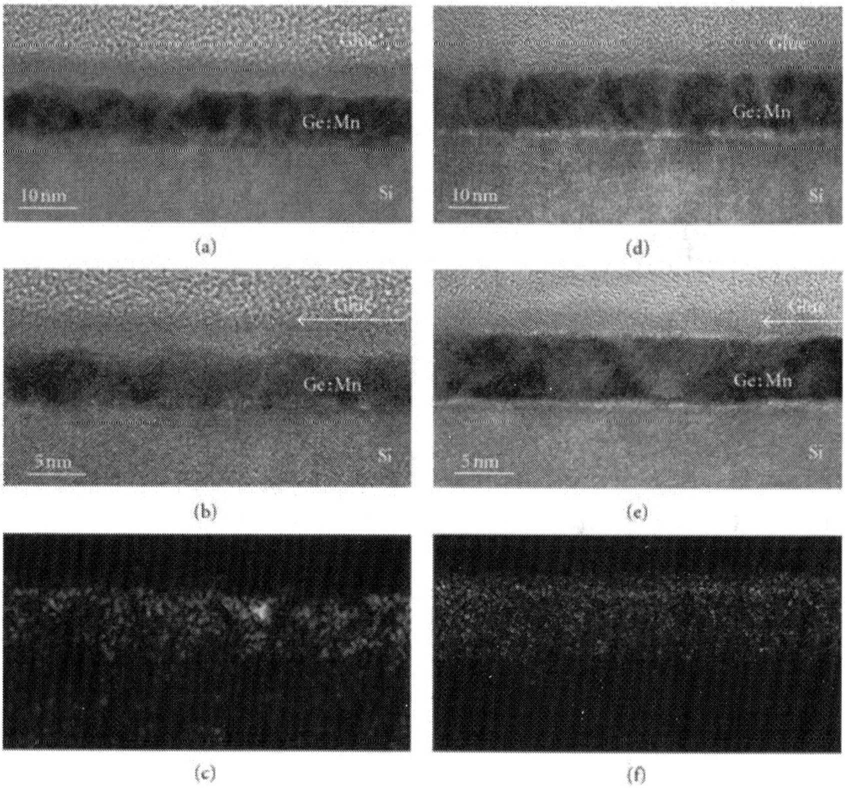

Figure 4. Typical TEM images of the as-grown (a) and the annealed (d) MnGe thin films with a thickness of 15 nm; the left column is for the as-grown film and the right column is the film after 400°C annealing. (b) and (e) are the corresponding HRTEM images of (a) and (d), respectively; (c) and (f) EFTEM elemental maps of Mn correspond to the regions given by (a) and (d), respectively. The white arrows in (b) and (e) show the separation between the glue and the amorphous MnGe layer.

To understand the structural variation for the two cases, high-resolution TEM was carried out and typical ⟨110⟩ zone-axis HR-TEM images are shown in Figures 4(b) and 4(e) for the two cases, respectively, where the amorphous layers can be clearly seen. From these HR-TEM images, stacking faults in the crystalline layers can also be seen in both cases. No structural clusters were found through our extensive HR-TEM investigations and no secondary phases were detected by our XRD measurements (not shown here). To further understand the Mn distribution in the Mn_xGe_{1-x} film (crystalline and amorphous), energy-filtered TEM was preformed. Figures 4(c) and 4(f) give the EF-TEM elemental maps of Mn that correspond to Figures 4(a) and 4(d), respectively. It is of interest to note that, in the as-grown case, a higher Mn concentration is found in the amorphous layer when comparing Figures 4(a) with 4(c). The overall concentration of Mn is low, so that the contrast in the Mn maps tends to be faint. Nevertheless, the distribution of Mn in the entire film can be clearly observed. In contrast, the Mn map (Figure 4(f)) in the annealed case shows some Mn distribution in the entire film, suggesting that the annealing has not only promoted crystallization, but also redistribute Mn in the entire Mn_xGe_{1-x} film. Nevertheless, a higher Mn concentration can still be observed in the surface region in the annealed sample (refer to Figure 4(f)). The fact that, for the as-grown sample, most of Mn was found in the topmost layer suggests that, during the growth of the Mn_xGe_{1-x} film, a dynamic Mn diffusion has taken place, even at a relatively low growth temperature of 250°C as in our case; Mn tends to diffuse towards the top surface.

Figure 5(a) shows the hysteresis loops at 298 K and 10 K. A clear hysteresis at 298 K indicates room-temperature ferromagnetism. Figure 5(b) shows that the plot of the magnetic moment as a function of temperature at a magnetic field of 100 Oe. The moment in the FC curve decreases very marginally as the temperature increases from 10 to 298 K, showing that the Tc is above the room temperature. The separation of the ZFC data from that of the FC shows a spin glass phase. Before and after annealing, there is no clear difference observed in magnetic properties. The Tc can only be inferred to be above 400 K as shown in Figure5(b).

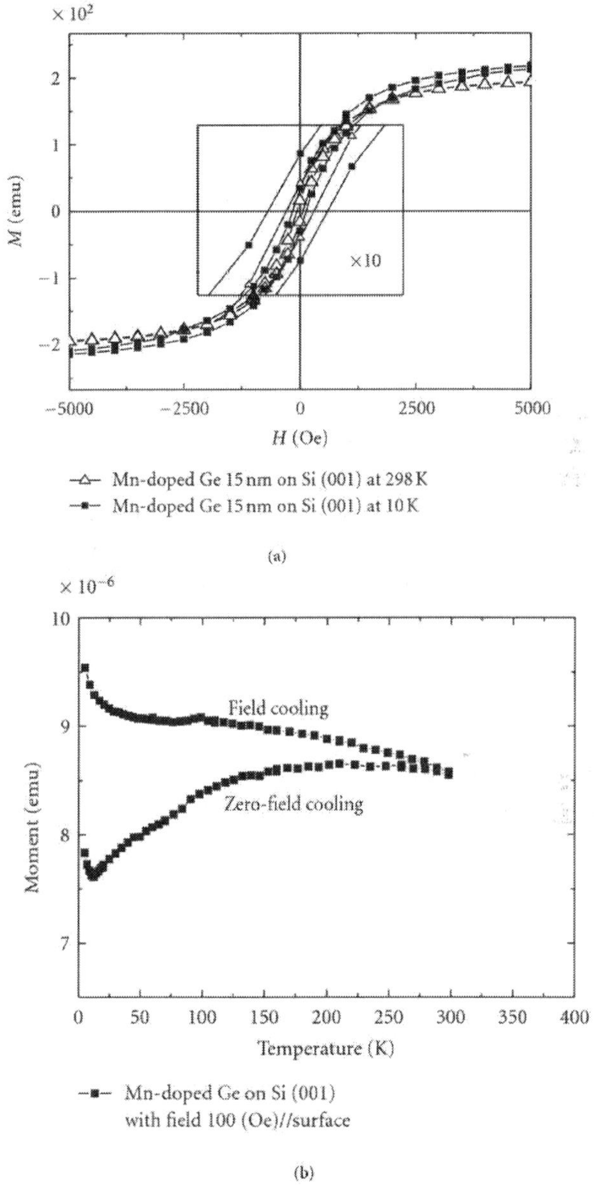

Figure 5. SQUID measurement of 4% Mn-doped Ge thin film grown on Si with a thickness of 15 nm. (a) Hysteresis loops measured at 10 and 298 K; (b) magnetic moment versus temperature measured with a magnetic field of 100 Oe.

The difference of structural and magnetic properties between the 80 nm thick and the 15 nm thick Mn_xGe_{1-x}thin films suggest that when it comes to the nanostructures, the strain and the Mn distribution could be different from that of the bulk films. There might be a threshold thickness for a certain Mn concentration, beyond which the nucleation of metallic clusters (such as Mn_5Ge_3 and $Mn_{11}Ge_8$) takes place and accounts for the RT ferromagnetism (Tc=296~300 K) [84, 108]. It is therefore of great interest to further investigate Mn_xGe_{1-x} nanostructures to explore their magnetic properties. Mn_xGe_{1-x} nanodots may be formed by ion implantation process of Mn into Ge and by a self-assembling process with MBE. Since the Mn ion implantation induces metallic clusters and defects [92], we will only focus on the Mn_xGe_{1-x} DMS QDs fabricated via MBE.

Self-Assembled DMS Mn_xGe_{1-x} QDs

$Mn_{0.05}Ge_{0.95}$ QDs were grown on p-type Si substrates. Cross-section TEM was carried out to determine the structural characteristics and the Mn composition. HR-TEM image reveals that a QD has a dome shape with a base diameter of about 30 nm and a height of about 8 nm (Figure 6(a)) The interface between the dot and the Si substrate has excellent lattice coherence. A careful inspection reveals that the dot is single crystalline without evidence of pronounced dislocations or stacking faults. However, because of the heavy Mn doping it is possible that some amount of point defects (such as Mn interstitials) may be present inside the dot, which is beyond the detect capability of conventional TEM.

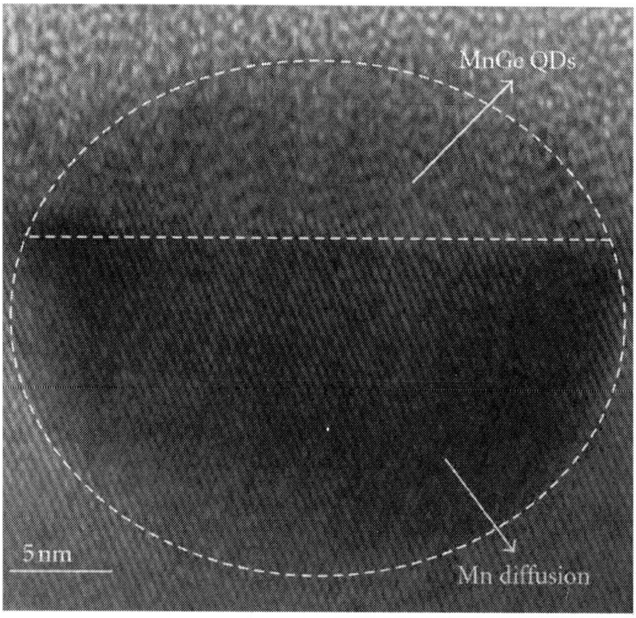

(a)

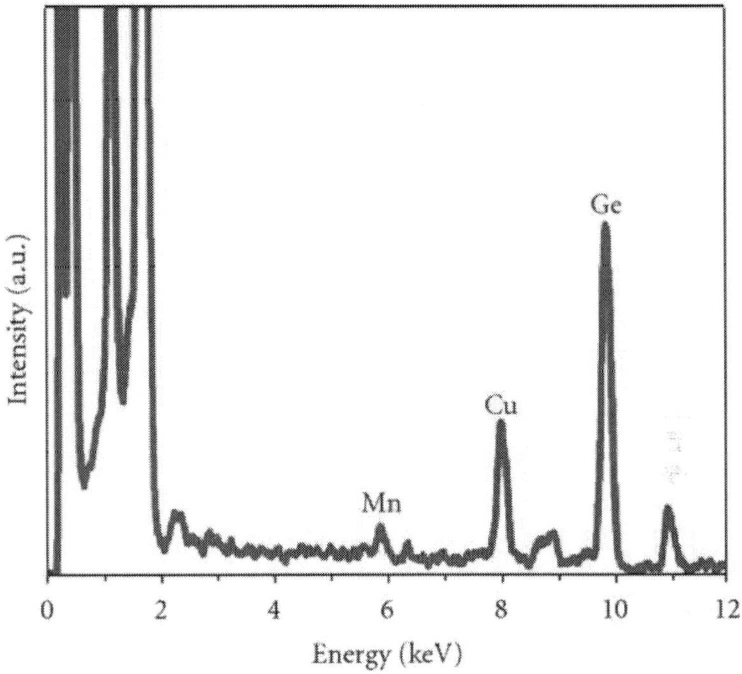

(b)

Figure 6. Structural properties of $Mn_{0.05}Ge_{0.95}$ quantum dots grown on a p-type Si substrate. (a) A high-resolution TEM cross-section image of typical $Mn_{0.05}Ge_{0.95}$quantum dot showing the detailed lattice structure. (b) Typical EDS spectrum showing that both Mn and Ge are present in the $Mn_{0.05}Ge_{0.95}$ quantum dot. The average Mn concentration in the quantum dots was estimated to be 4.8 ± 0.5%.

Directly underneath the $Mn_{0.05}Ge_{0.95}$ QD, Mn diffuses into Si substrate and forms a strained MnSi area, which has the same diameter as the top $Mn_{0.05}Ge_{0.95}$ QD, but a height of about 16 nm (Figure 6(a)). This diffusion behavior is not unusual as it was also observed in (In, Mn)As and (In, Cr)As QDs systems [120, 121]. However, the migrations of Mn into the substrate make it difficult to accurately determine Mn concentration inside the dot. To address this challenge, we have performed energy dispersive X-ray spectroscopy (EDS) experiments (in a scanning TEM mode) to analyze the Mn composition at nanoscale (Figure 6(b)). When the EDS was performed, electron probes with several nanometers in diameter were illustrated on the $Mn_{0.05}Ge_{0.95}$ QDs and their underlying Si substrate. Consequently, Si peaks are constantly observed in the EDS spectra. To estimate the Mn concentrations, we performed a quantitative analysis of atomic percentages of Mn and Ge and artificially discounted the Si peak.

The EDS analysis over many QDs reveals a Mn/Ge atomic ratio of about 0.144 : 1. Since approximately 1/3 volume fraction of Mn is distributed in the Mn_xGe_{1-x} QD [40], the average Mn concentration can be estimated to be 4.8 ± 0.5%. Note that the deviation was determined by a thorough study of many $Mn_{0.05}Ge_{0.95}$ QDs. Both the HR-TEM investigations and the composition analysis suggest that each individual QD is a single crystalline DMS system.

Magnetic properties were studied using a superconducting quantum interference device (SQUID) magnetometer at various temperatures. Figure 7(a) and two corresponding insets show the temperature-dependent hysteresis loops when the external magnetic field is parallel to the sample surface (in-plane). The field-dependent magnetization indicates a strong ferromagnetism above 400 K. The saturation magnetic moment per Mn atom is roughly estimated to be 1.8 μ_B at 5 K. A fraction of roughly 60% of Mn is estimated to be activated assuming that each Mn has a moment of 3 μ_B [27, 58]. The Arrott plots were also made in order to evaluate the *T2c* (Figure 7(b)). By neglecting the high-order terms, the magnetic field can be expressed in the following equation [122]:1

$$H = \frac{1}{\chi}M + \beta M^3 \quad \text{or} \quad \frac{H}{M} = \frac{1}{\chi} + \beta M^2, \tag{1}$$

where *H* is the external magnetic field, *M* is the magnetic moment from the sample, χ is the susceptibility, andβ is a material dependent constant.

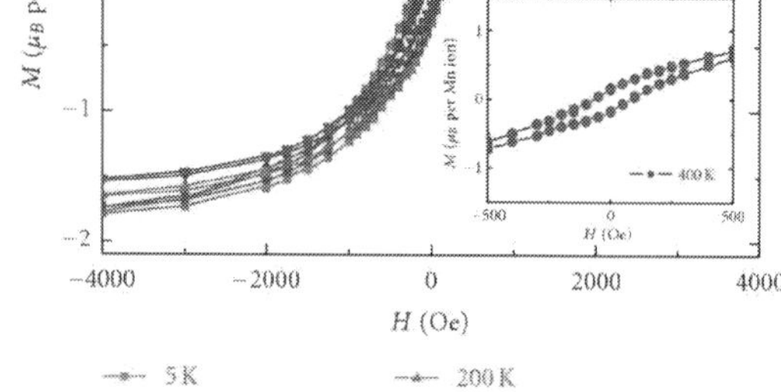

(a)

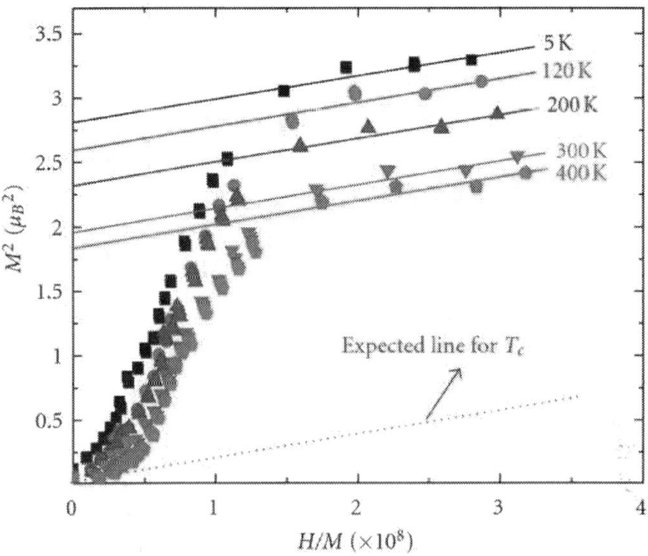
5 K
120 K
200 K
300 K
400 K
Expected line for T_c
M^2 (μ_B^2)
H/M ($\times 10^8$)

(b)

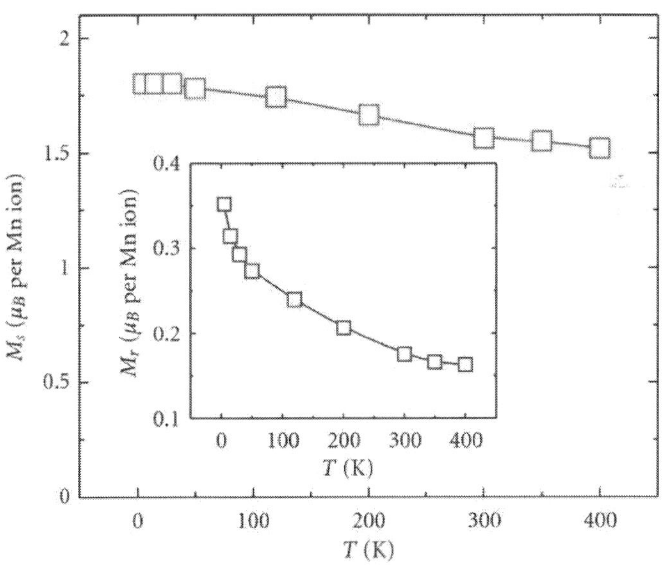
M_s (μ_B per Mn ion)
M_r (μ_B per Mn ion)
T (K)
T (K)

(c)

(c)

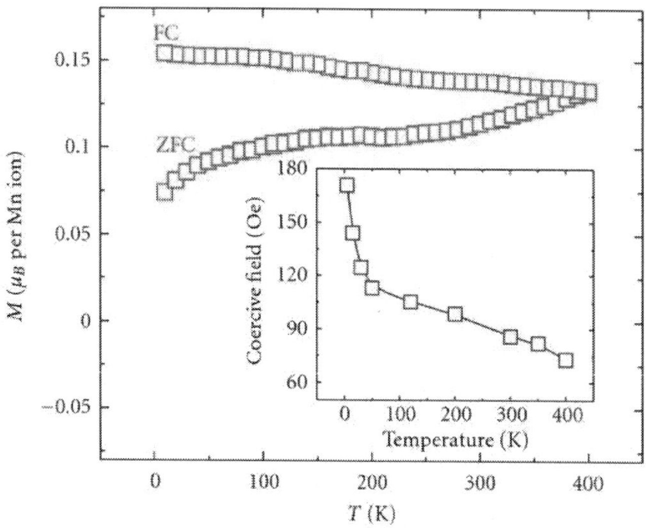

(d)

Figure 7. Magnetic properties of the $Mn_{0.05}Ge_{0.95}$ quantum dots grown on a p-type Si substrate. (a) Hysteresis loops measured at different temperatures from 5 to 400 K. (b) The Arrott plots were made to obtain the Curie temperature. Consistent with (a), the Curie temperature is projected to be above 400 K; (c) The temperature dependence of saturation moments. The inset gives the remnant moments with respect to temperature; (d) Zero-field cooled and field cooled magnetizations of quantum dots with a magnetic field of 100 Oe; the inset shows the coercivity values at different temperatures. The external magnetic field is in parallel with the sample surface.

According to (1), $M2$ can be plotted as a function of H/M. When H/M is extrapolated to $M2=0$, the intercept on the H/M axis gives $1/\chi$. The Tc for this material can be obtained when $1/\chi$ vanishes. Figure 7(b) shows the Arrott plots at several temperatures. It can be seen that even at 400 K the intercept ($1/\chi$) on the H/M axis does not vanish, which means that the susceptibility still has a finite value and the Tc has not been reached yet. By using the slope obtained at 400 K, a dashed line can be drawn as shown in Figure 7(b), in which a Tc is projected to be beyond 400 K. This is in a good agreement with the data from the hysteresis loops showing the magnetic order above 400 K. Figure 7(c) and corresponding inset show the temperature-dependent saturation and remnant moments per Mn ion, respectively. Both of them demonstrate weak temperature dependences and substantial amount of magnetization moments remain even at 400 K.

ZFC and FC magnetizations were measured with a magnetic field of 100 Oe as shown in Figure 7(d). The magnetic moments do not drop to zero, suggesting a high Tc beyond 400 K, which is in a good agreement with the Arrott plots in Figure 7(b). From these two curves, one can also infer the formation of a single phase in this material system, that is, DMS QDs. The wide separation of the ZFC and FC curves in the temperature range of 5 to 400 K shows the irreversibility of susceptibilities, possibly arising from strain-induced anisotropy as a large lattice mismatch exists between Si and Ge [61]. The temperature-dependent coercivity is shown in Figure 7(d)inset. The coercivity decreases from 170 Oe (at 5 K) to 73 Oe (at 400 K). The small coercivity in the entire temperature range measured features a soft ferromagnetism which originates from Mn ions diluted in the Ge matrix [30]. The above magnetic properties support the fact that the $Mn_{0.05}Ge_{0.95}$ QDs exhibit a DMS-type ferromagnetic order.

It is well known that the direct interaction (superexchange) between d-shells of the adjacent Mn atoms leads to an antiferromagnetic configuration of the d-shell spins [8]. Therefore, the presence of Mn-Mn clusters would lead to the antiferromagnetism; but when this pair is surrounded by other Mn atoms, the ferri- or ferromagnetic coupling may appear. However, the extensive high-resolution TEM did not reveal observable Mn clusters although such a possibility cannot be completely excluded (Figure 6). Thus, the origin of strong ferromagnetism possibly comes from the $Mn_{0.05}Ge_{0.95}$ QDs via the hole mediation process. In general, when Mn dopants are introduced into Ge, they produce two acceptor levels, that is, 0.16 eV and 0.37 eV above the valence band [123]. When the doping concentration becomes high, the impurity bands could be developed (instead of being separate acceptor levels). However, theoretically the formation of such impurity bands may not affect the physics of ferromagnetism [124]. In the following, we will analyze the Mn-doping behavior by considering a single acceptor level of 0.16 eV in an approximate form. At a sufficiently high temperature, Mn acceptors can be activated to generate holes. These holes are itinerant among a number of Mn ions and align Mn local spins along one direction via the p-d exchange coupling to reduce the total energy of the system, resulting in the hole-mediated ferromagnetism [125]. Experimentally this phenomenon has been widely observed in several material systems, such as group III-V (GaAs and InAs) [2, 126], group II-VI (ZnO) [127], and group IV (Ge) [27, 39]. While most of research focuses on bulk DMS, of particular interest is to study the ferromagnetism origin in quantum structures such as (In, Mn)As and Mn_xGe_{1-x} QDs. Previously, Chen et al. [39]and others [120] have suggested that the carrier confinement in a DMS QD can strengthen hole localization and subsequently enhance the thermal stability of magnetic polarons giving rise to a higher Tc than bulk films.

Atomic force microscopy (AFM) and magnetic force microscopy (MFM) measurements were carried out to investigate the morphology and ferromagnetism of $Mn_{0.05}Ge_{0.95}$ QDs at 320 K, respectively. The average dot size is 50 nm in base diameter and 6 nm in height. The dot density is about 6×10^9 cm^{-2} (Figure 8(a)). The corresponding MFM image was taken by lifting up the MFM probe 25 nm above the topographic height of the sample in a phase detection mode (Figure 8(b)). The appearance of bright-and-dark areas in the MFM image clearly shows the formation of magnetic domains in the $Mn_{0.05}Ge_{0.95}$ QDs, which is similar to (In, Mn)As DMS QDs [128]. Figures 8(c)–8(f) show enlarged MFM images of several individual $Mn_{0.05}Ge_{0.95}$ QDs. By reversing the tip magnetization, opposite contrast was observed for each dot, indicating that the magnetic signals originated from the top $Mn_{0.05}Ge_{0.95}$ QDs. Note that each QD is a single domain "particle." During the magnetization process, the domain would rotate preferentially along the magnetic field to produce net magnetization moments. Since the experiments were performed at 320 K, the formation of metallic phases such as Mn_5Ge_3 and $Mn_{11}Ge_8$ can be easily ruled out because they have low Curie temperatures of 296~300 K. Overall, the above MFM results agree well with the TEM observations and the ferromagnetic order at high temperature obtained in the SQUID measurements.

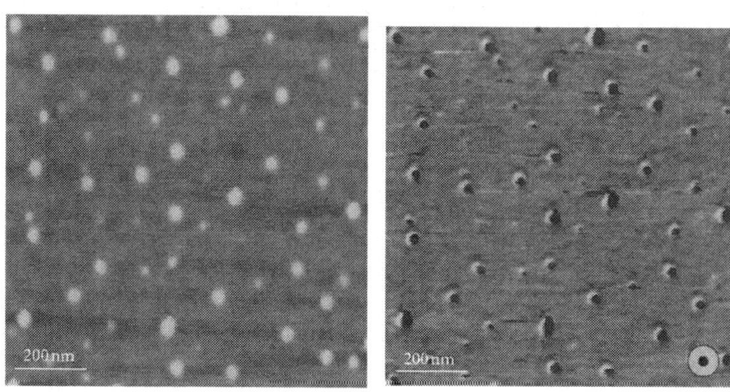

(a) (b)

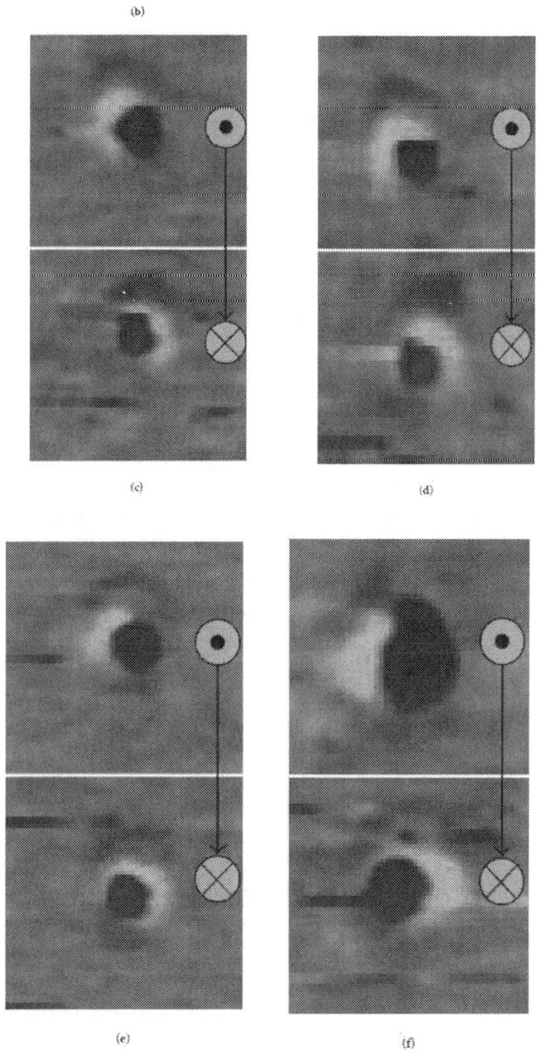

Figure 8. AFM and MFM images of the $Mn_{0.05}Ge_{0.95}$ quantum dots measured at 320 K. (a) Typical AFM image of $Mn_{0.05}Ge_{0.95}$ quantum dots; (b) corresponding MFM image with the tip magnetization pointing toward the sample; (c)~(f) enlarged MFM images of individual quantum dots taken from (b). From these MFM measurements, opposite contrasts were observed when applying an opposite magnetization to the tip. The results suggest that the magnetic signals come from the $Mn_{0.05}Ge_{0.95}$ quantum dots. The formation of Mn_5Ge_3 and $Mn_{11}Ge_8$ can be excluded because of their low Curie temperatures of 296~300 K.

Discussions

We have reviewed the growth of $Mn_{0.05}Ge_{0.95}$ by MBE, ion implantation, and bulk sintering, emphasizing the growth temperature effect on the phase formation and magnetic properties. While the current motivation is driven by the prospect of enhancing Curie temperatures, the experiments show a strong tendency of metallic precipitates developed in bulk and relatively thick Mn_xGe_{1-x} crystals. The fundamental understanding here points to a spin-glasslike feature in Mn_xGe_{1-x} due to the formation of various lattice coherent nanostructures and metallic precipitates, which exhibit magnetic blocking behaviors under different temperature regimes. The challenge to eliminate these clusters seems to be overwhelming and nearly impossible. However, when the material dimension decreases, it evidences a change of material structures being free of precipitates and a significant increase of Curie temperatures over 400 K. The preliminary explanation on these behaviors can be attributed to the carrier confinement in a DMS QD which strengthens hole localization and subsequently enhances the thermal stability of magnetic polarons giving rise to a higher Tc than those of bulk films. Nevertheless, the experimental data suggest that with nanoscale structures, the quantum confinement effect comes into being which significantly influences the exchange coupling between the confined holes and the localized Mn^{2+}. Future progress in this direction will involve a detailed theoretical treatise to understand the quantum confinement effect on magnetic properties in a quantitative picture.

ELECTRIC-FIELD-CONTROLLED FERROMAGNETISM

Introduction

Electric field control of ferromagnetism has a potential to realize spin field-effect transistors (spin FETs) and nonvolatile spin logic devices via carrier-mediated effect [129, 130]. With the manipulation of carrier spins, a new generation of nonvolatile (green) computing systems could be eventually developed for many low-power-dissipation applications in all fields, including sensor network, health monitoring, information, and sustainable wireless system. Since Datta and Das [131] first introduced the concept of spin FETs in 1990, enormous efforts were dedicated to creating a device wherein the carrier transport is modulated by electrostatic control of carrier spins [2, 20, 22, 23, 132–136]. To understand and exploit this controllability, several theoretical models were proposed to explain the ferromagnetic coupling in DMS on a microscopic scale (also elaborated in the theoretical section): the Zener Kinetic-exchange [10], double-exchange [137], and the RKKY interaction [138, 139]. These models share a common feature that a spontaneous ferromagnetic order is carrier-mediated through the increase of carrier concentrations. One of major challenges, however, is to search an ideal material with room-temperature controllable spin states [21, 22, 140]. In recent years, emerged DMSs became one of the promising candidates since they could possibly offer high Tc in excess of 300 K [137]. The demonstration of the carrier-medicated ferromagnetism involving correlated electron/hole systems leads to a para- to ferromagnetism phase transition [27, 39, 130, 137]. In principle, the collective alignment of spin states in these DMSs can be manipulated by the modulation of carrier concentrations through gate biasing in a FET structure [2, 140]. For this kind of spin FETs, the "source" and "drain" may be completed through "nanomagnet," which are in turn controlled by the gate; and no carrier transport is needed. Clearly, one may also involve the control of source-drain conductance by gate-voltage-induced precession of injected spins (from the source). Since early 2000s, a significant progress was achieved [2, 23, 130, 135], in which the ferromagnetism of a (In, Mn)As channel layer could be effectively turned on and off via electric fields in a gated FET. Such extraordinary field-modulated ferromagnetism immediately rendered the development of future spintronic devices. However, the manipulation of ferromagnetism was limited because of low Tc of the Mn-doped III-V materials [141]. Therefore, a search for new DMS materials with Tc>300 K and carrier-mediated ferromagnetism becomes a current global challenge [10, 140].

Electric-Field-Controlled Ferromagnetism in Mn-Implanted Ge Dots

In parallel with the experimental efforts on field controlled ferromagnetism in III-V material system, we have previously investigated the magnetic properties of Mn_xGe_{1-x} nanostructures by ion implantation of Mn [39]. We demonstrated the hole-mediated effect in Mn-implanted nanodot structures and the modulation of their ferromagnetism by applying gate biases in MOS capacitors at a low temperature of 10 K. Figure 9 shows the magnetic properties of samples with Mn implantation into Ge via the nanopatterned SiO_2 mask on different doped substrates. It is observed that the Mn-implanted p+ Ge (10^{19} cm^{-3}) has stronger ferromagnetic signals than that n-type Ge (10^{17} cm^{-3}), which may indicate a hole-mediated effect. Note that we cannot completely exclude the possibility of the formation of dopant complexes, such as B-Mn and Sb-Mn in the p+ and n-type Ge, respectively. Taking advantage of this effect, we further fabricated a MOS capacitor device with Mn-implanted Ge as the channel layer. Figure 10 shows the magnetic moment versus applied field for the MOS structure on the n-type substrate under various gate biases at 10 K. When the voltage is applied from −6 to −30 V, the hole concentration in the Mn_xGe_{1-x} channel increases and the magnetic hysteresis becomes larger. At 0 V bias, only a line with a small slope and a small negligible loop are seen; this is because a low density of excess holes is present. When a positive voltage is applied to the gate, holes are partially depleted, resulting in a decrease of the channel magnetization and in the disappearance of magnetic hysteresis. These experimental results demonstrated the hole-mediated effect in Mn_xGe_{1-x}, consistent with the finding by Park et al. [27]. However, the metallic precipitates such as Mn_5Ge_3 and $Mn_{11}Ge_8$ and implantation damages were found in these nanostructures, which made the system rather complex and might have partly jeopardized the hole-mediated effect.

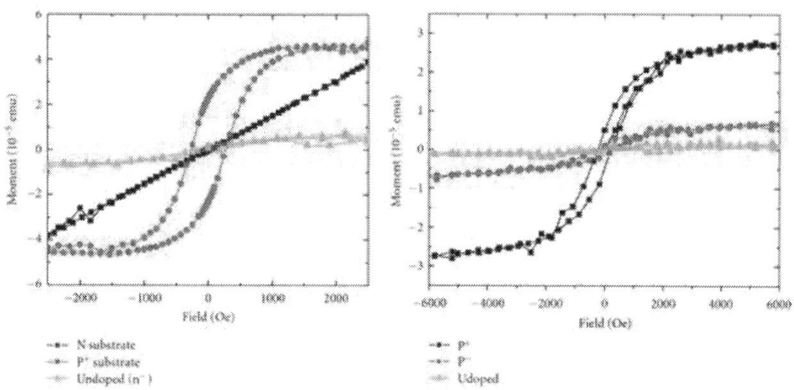

Figure 9. The magnetic properties of Mn-implanted Ge via a nanopatterned SiO_2 mask. The hole-mediated effect can be clearly observed in the Mn-implanted p-type Ge sample.

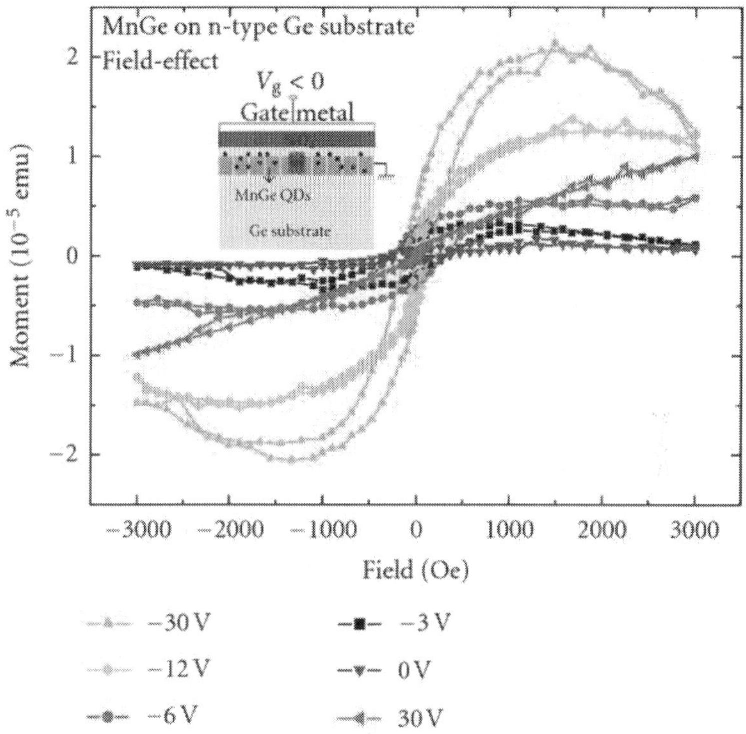

Figure 10. Magnetic moment versus applied field under various biases from −30 to 30 V at 10 K. The obtained hysteresis is the largest under −30 V gate voltage, and the ferromagnetic phase is turned off under +30 V gate bias. The inset shows the spin-gated structure used to investigate C-V and gate-bias-dependent magnetic properties.

Electric-Field-Controlled Ferromagnetism in Self-Assembled DMS $Mn_{0.05}Ge_{0.95}$ QDs

Since the self-assembled $Mn_{0.05}Ge_{0.95}$ QDs have a high Tc above 400 K and are free of metallic phases ($Mn_5Ge_3$and $Mn_{11}Ge_8$), it can be potentially used as a channel layer in a MOS device to study the modulation of ferromagnetism by electric field. The device structure consists of a metal gate (Au, 200 nm), Al_2O_3 (40 nm), $Mn_{0.05}Ge_{0.95}$ QDs (6 nm), a wetting layer (<0.6 nm), a p-type Si substrate (1×1018 cm^{-3}), and a back metal contact (Au, 200 nm). A relative thick dielectric of Al_2O_3 (40 nm) is employed to ensure a small leakage current to a level below 10^{-6} A/cm^2. Figure 11 shows the electric field controlled ferromagnetism performed at 50, 77, and 100 K, corresponding to (a)–(c), (d)–(f), and (g)–(i), respectively. Due to the similarity of the data, we take 77 K as an example to describe the device operation in Figures 11(d)–11(f).

Figures 11(d) and 11(e) show the hysteresis loops by SQUID with negative and positive biases on the MOS gate at 77 K, respectively. Under a negative bias, the holes are attracted into the channel of the device (accumulation). In this circumstance, however, the hysteresis loop does not show a remarkable change (Figure 11(d)). This could be explained by the fact that even at 0 V, the QDs device is already accumulated with enough holes to induce ferromagnetism. In other words, the hole-mediated effect is sufficient to align a majority of the activated Mn ions along one direction in each individual QD. Further increasing negative bias does not change much on the hole concentrations. On the contrary, with the positive bias, a large amount of holes are depleted into the p-type Si so that hole-mediated effect is notably reduced. In fact, the surface of the device can have a high concentration of electrons if the leakage is limited (inversion). As a result, the Mn ions start to misalign because of the lack of holes. The saturation moment per Mn ion decreases more than 10 times as the gate bias increases from 0 to +40 V (Figure 11(e)). It should be noted that, at +40 V, the saturation and remnant moments of the $Mn_{0.05}Ge_{0.95}$ QDs become fairly weak, resembling a "paramagnetic-like" state. Figure 11(f) summarizes the change of remnant moments as a function of gate voltage. A similar trend in the carrier density as a function of the gate bias was observed from CV curves (not shown here), suggesting a strong correlation between hole concentrations and the ferromagnetism, that is, the hole-mediated effect. The inset in Figure 11(f) displays an enlarged picture to clearly show the change of remnant moments with respect to the gate bias. By increasing the measurement temperature to 100 K (Figures 11(g)–11(i)), the modulation of the ferromagnetism becomes less pronounced compared to those at 50 and 77 K due to the increased leakage current in our MOS devices. The above results evidently demonstrate that the hole-mediated effect does exist in this material system.

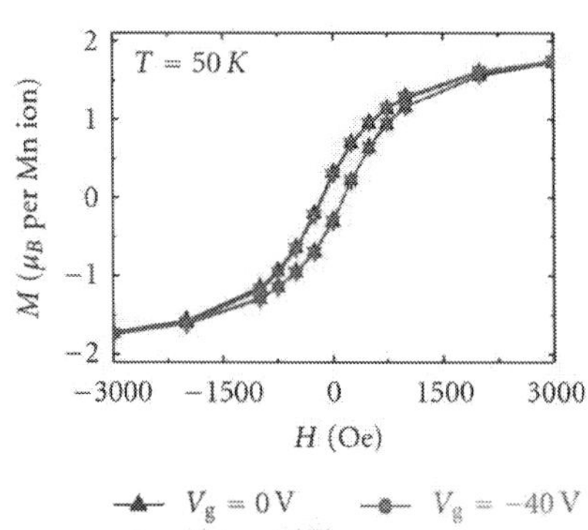

(a)

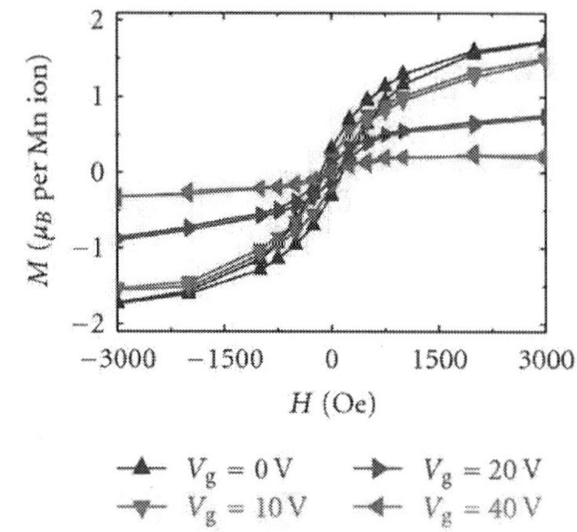
M (μB per Mn ion)
2
1
0
−1
−2
−3000
−1500
0
1500
3000
H (Oe)
Vg = 0 V
Vg = 10 V
Vg = 20 V
Vg = 40 V

(b)

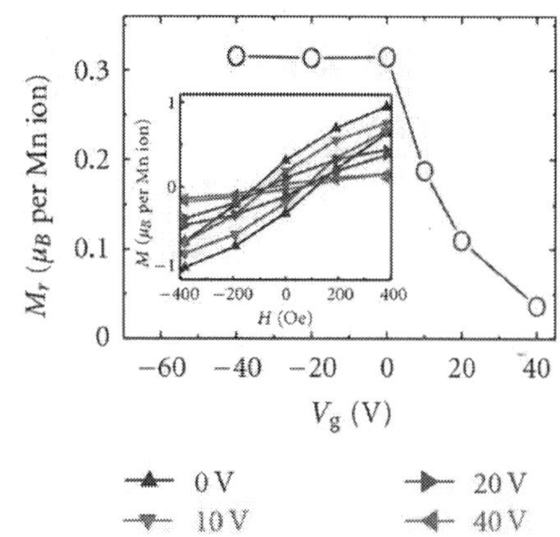
Mr (μB per Mn ion)
0.3
0.2
0.1
0
−60
−40
−20
0
20
40
Vg (V)
M (μB per Mn ion)
H (Oe)
−400
−200
200
400
0 V
10 V
20 V
40 V

(c)

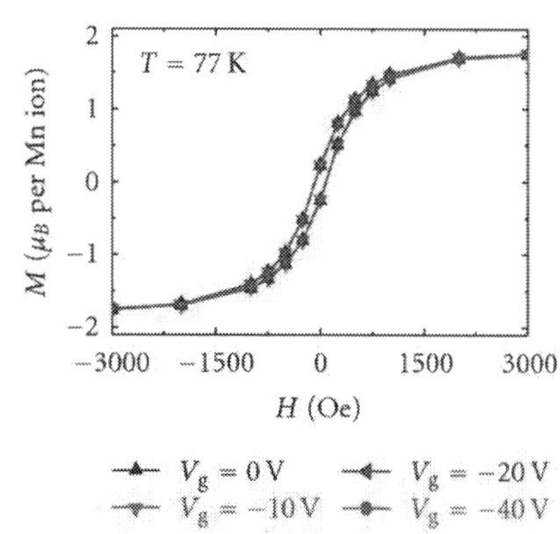
T = 77 K
M (μB per Mn ion)
2
1
0
−1
−2
−3000
−1500
0
1500
3000
H (Oe)
Vg = 0 V
Vg = −10 V
Vg = −20 V
Vg = −40 V

(d)

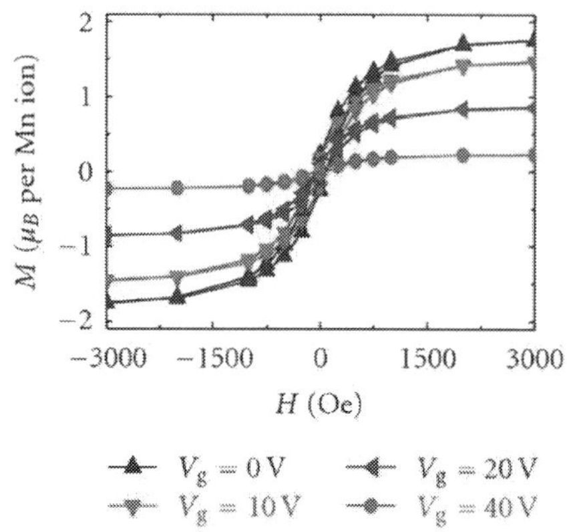
M (μ_B per Mn ion)
H (Oe)
V_g = 0 V
V_g = 20 V
V_g = 10 V
V_g = 40 V

(e)

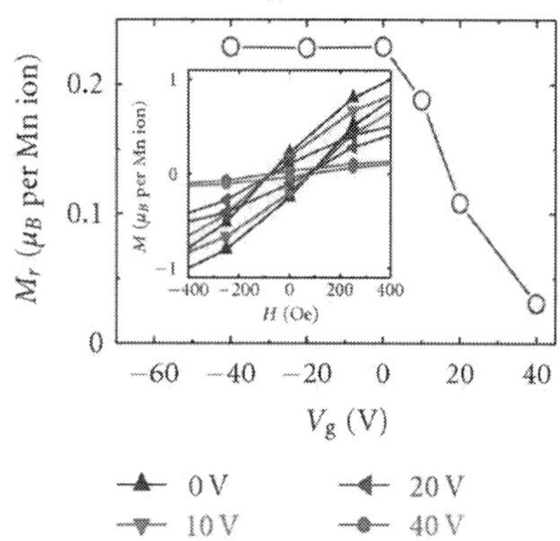
M_r (μ_B per Mn ion)
V_g (V)
M (μ_B per Mn ion)
H (Oe)
0 V
20 V
10 V
40 V

(f)

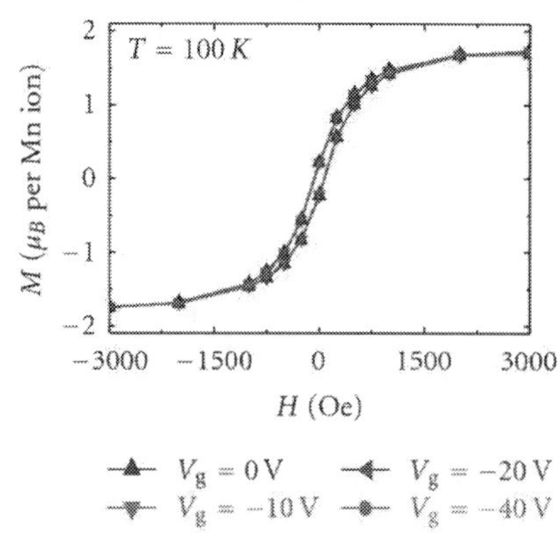
T = 100 K
M (μ_B per Mn ion)
H (Oe)
V_g = 0 V
V_g = −20 V
V_g = −10 V
V_g = −40 V

(g)

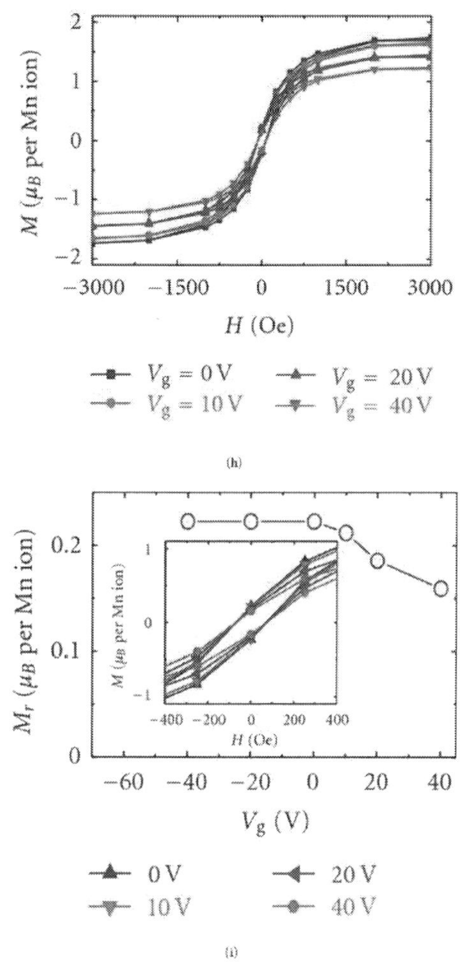

Figure 11. Control of ferromagnetism of $Mn_{0.05}Ge_{0.95}$ quantum dots by applying electric field at 50 K ((a)~(c)), 77 K ((d)~(f)), and 100 K ((g)~(i)). (a), (d), and (g) Hysteresis loops with zero and negative bias of −10, −20 and −40 V on the gate; (b), (e), and (h) the hysteresis loops with zero and positive bias of +10, +20, and +40 V; (c), (f), and (i) remnant moments with respect to the gate bias. Insets of (c), (f) and (i) are enlarged figures from the central part of (b), (e), and (h) to clearly show the change of remnant moments, respectively. It is found that both the saturation and remnant moments can be manipulated by applying biases on the MOS gate at 50, 77, and 100 K.

Discussions

Electrical-field-controlled ferromagnetism has been successfully demonstrated in Mn-doped III-V DMSs [2,23, 130, 135, 140]. It is well established that the variation of hole concentrations renders such a controllability [2]. The low Tc of III-V DMS, however, presents a challenge to further realize room-temperature-controlled ferromagnetism [141]. Alternatively, the recent experiments on $Mn_{0.05}Ge_{0.95}$ QDs show a high Tc above room temperature and gate-modulated ferromagnetism over 100 K [40]. While the leakage current suppressed the gate modulation, room-temperature-controlled ferromagnetism would not be impossible because of the high Tc of this system. Research in this direction may open up a pathway for achieving room-temperature Ge-based (and other) spin-FETs and spin logic devices. These spintronic devices could potentially replace the conventional FETs with lower power consumptions and provide additional new device functionalities, which is beyond today's mainstream CMOS technology of microelectronics.

METALLIC MN_XGE_{1-X} (X ~ 20%) NANODOT ARRAYS

Introduction

There is also a need to develop ferromagnet/semiconductor hybrid structures for semiconductor spintronics since they have magnetic and spin-related functions and excellent compatibility with semiconductor device structures [142, 143]. By embedding magnetic nanocrystals into conventional semiconductors, a unique hybrid system can be developed, allowing not only utilizing the charge properties but also the spin of carriers, which immediately promises next-generation nonvolatile magnetic memories and sensors [144, 145]. On the other hand, spin injections into the semiconductor can be dramatically enhanced via coherent nanostructures, which considerably reduce undesired spin scatterings [146]. Although magnetic hybrid systems, such as MnAs/GaAs, have been extensively studied over several decades, the control (over the spatial location, shape and geometrical configuration) of the magnetic nanostructures (for instance MnAs) still remains a major challenge to further improve the performance of the related magnetic tunnel junctions (MTJs) and spin valves [147].

Growth Methods

We employed a concept of stacked Mn_xGe_{1-x} (x ~ 20%) (hereafter, we denote it as MnGe) nanodots by alternatively growing MnGe and Ge layers with designated thicknesses (nominal 3 nm thick MnGe and 11 nm thick Ge). It is well known that Mn doping in Ge induces compressive strain because of its larger atomic size [148], assuming that no lattice defects are generated during the doping process, that is, lattice coherence. Mn-rich MnGe nanodots induced by the spinodal decomposition should be strained if the lattice coherence between the nanodots and the matrix remains. Once the strained nanodots are developed, a thin Ge spacer layer, subsequently deposited with an optimized thickness, will retain the perfect lattice coherence with the underneath nanodots. This enables the existing nanodots to exert strain on the Ge spacer layer and produce "strained spots," which, in turn, become preferred nucleation sites for successive nanodots. Eventually, multilayered and vertically aligned MnGe nanodot arrays can be produced, similar to the scenarios of stacked InAs/GaAs [149, 150] and Ge/Si [151] quantum dots. Indeed, by employing this innovative approach, we achieved the growth of coherent self-assembled MnGe nanodot arrays with an estimated density of 10^{11}~10^{12} cm^{-2} within each MnGe layer.

Ten periods of MnGe nanodots were epitaxially grown on Ge (100) and GaAs (100) substrates by the MBE system. TEM and EDS in the scanning TEM (STEM) mode were performed to understand the nanostructures and compositional variations of the resulting thin films. Figures 12(a) and 12(c) are typical plane-view and cross-sectional TEM images and show the general morphology of the MnGe nanostructures, viewed along the ⟨100⟩ and ⟨011⟩ directions, respectively. A high density of dark nanodots can be clearly seen in both cases. Based on the magnified cross-sectional image shown in Figure 12(d), the nanodot arrays are clearly observed with 10 stacks along the growth direction although not perfectly vertical. In order to determine the composition of the dark dots, EDS analyses in the STEM mode were carried out and typical plane-view and cross-sectional STEM images are shown in Figures 12(b) and 12(e), respectively. Figure 12(f) is the EDS result taken from a typical dot and shows clearly the Mn and Ge peaks. Figures 12(g) and 12(h) present EDS line scans using the Mn *K* peak for the dots marked by G and H in Figures 12(b) and 12(e), respectively, indicating high concentrations of Mn inside the dots. Taking all these comprehensive TEM results into account, it is concluded that the nanodots are Mn rich when compared with the surrounding matrix.

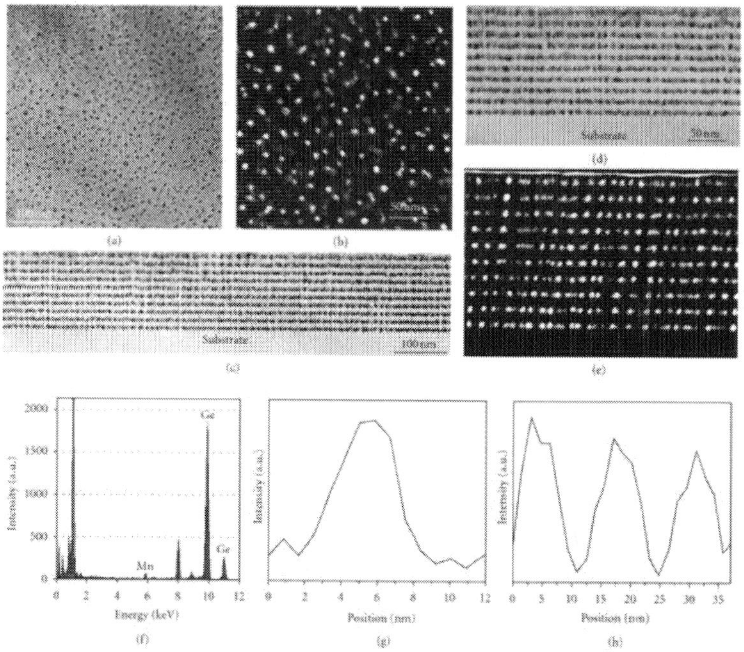

Figure 12. ransmission electron microscopy (TEM), scanning TEM, and energy dispersive X-ray spectroscopy (EDS) results of the multilayer MnGe nanodots. (a) A typical low-magnification plane-view bright-field TEM image showing MnGe nanodots (dark spots). (b) A plane-view low-angle dark-field STEM image showing the MnGe nanodots (white spots). (c) A low-magnification cross-sectional bright-field TEM image showing the obtained MnGe nanodot array in a large area. (d) A higher-magnification cross-sectional TEM image and (e) a cross-section STEM image, both showing the MnGe nanodot arrays. (f) An EDS profile showing the Mn and Ge peaks. (g, h) EDS line-scan profiles of the marked line in (b) and (e) using Mn *K* peak, respectively, confirming nanodots being Mn rich. All TEM images are taken from the same sample.

Figure 13(a) shows a high-magnification TEM image taken from a thin area, where several aligned MnGe nanodots can be evidently observed. The distance between two vertically adjacent nanodots (along the growth direction) is measured to be 14 ± 1 nm, well matched with the designed period of a 11 nm thick Ge spacer layer and a 3 nm thick MnGe layer. It should be noted that these nanodots are uniform in size with an elliptical shape (a dimension of 5.5 ± 0.5 nm and 8 ± 0.3 nm in the horizontal and vertical directions, resp.), as demonstrated in Figure 13(a). Since the nominal thickness of the MnGe layer (3 nm) is far less than the dot vertical dimension (8 nm), it suggests that, during the growth of the MnGe thin film, Mn not only diffuses laterally (to form dots), but also migrates vertically into the adjacent Ge spacer layers, primarily in the proximity of the dot regions, resulting in ellipse-shaped nanodots.

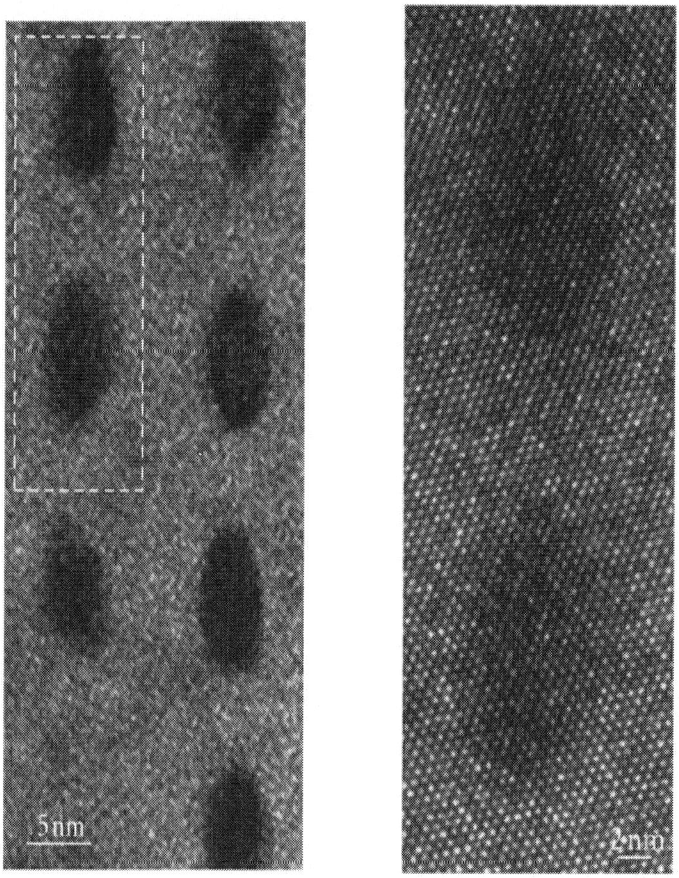

(a) (b)

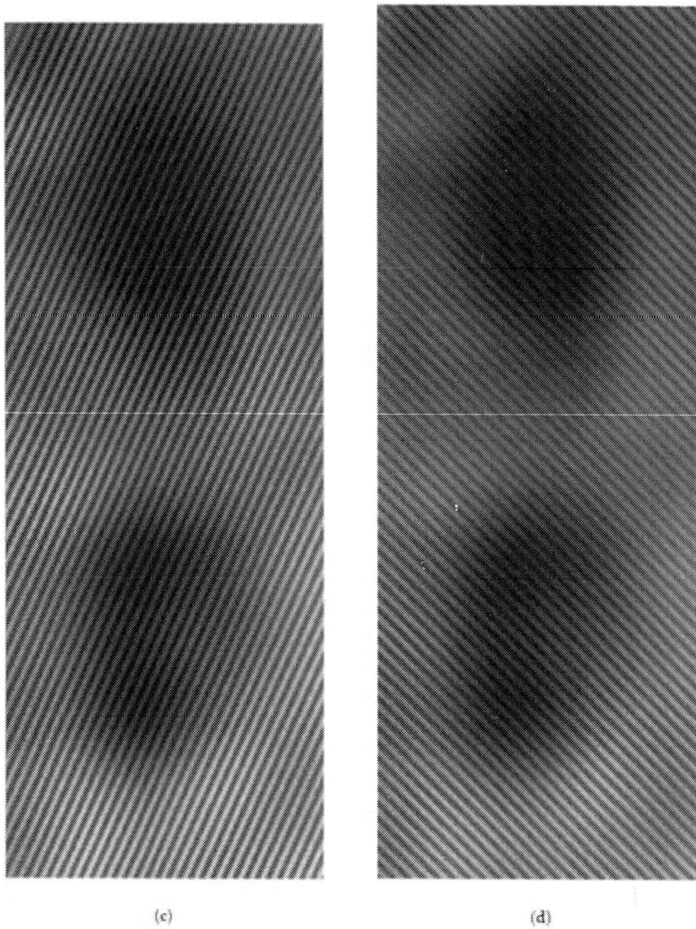

Figure 13. (a) A typical TEM image showing several vertically aligned MnGe nanodots. (b) The HRTEM images of the MnGe nanodots (the selected area in (a)) showing a perfect diamond structure as the Ge matrix. (c, d) Bragg filtering of ±(111) (c) and ±(111) (d) reflections, respectively, where no dislocation or distortion was observed. The dark contrast of the nanodots indicates the existence of significant strain.

Magnetotransport Properties of Nanodot Arrays

The resistivity measurements were carried out to probe the carrier transport under different temperatures. It was found that the temperature-dependent resistivities rapidly increase with decreasing temperature due to the carrier freeze-out effect at low temperatures, which is typically observed in doped semiconductors [152]. Considering the embedded MnGe nanodots, the rise in resistivities at low temperatures also suggests a strong localization of carriers, which takes place at the Mn sites and/or at the MnGe/Ge interfaces, similar to the scenario of MnSb clusters in InMnSb crystals [153]. The temperature-dependent resistivity can be generally described by [154]

$$\rho(T) = \rho_0 \exp\left[\left(\frac{T_0}{T}\right)^{1/\alpha}\right], \qquad (2)$$

where $\rho(T)$ is the temperature-dependent resistivity; $\rho 0$ and $T0$ denote material parameters, α is a dimensionality parameter: α=2. For one-dimensional (1D), α=3 for 2-d, and α=4 for 3-d systems. In order to reveal the carrier transport mechanisms at different temperature regions, fittings were performed in the plots of lnρ as a function of $T-\alpha$ (Figure 14(a)). The best fittings were found when α equals 1 and 4 in the high-temperature and low-temperature regions, respectively, corresponding to the carrier transport via the band conduction [155] (thermal activation of acceptors) and the 3-d Mott's variable range hopping processes [154]. According to the fitting results to (2), the obtained nanodot arrays show a dominated hopping process below 10 K. At such a low temperature, the majority of free holes are recaptured by the acceptors. As a result, the free-hole band conduction becomes less important and hole hopping directly between acceptors in the impurity band contributes mostly to the conductivity [155]. Above 100 K, the conduction is dominated by the thermal activation of the holes (the band conduction). A thermal activation energy ($E\alpha$) of 15 meV can be obtained from (2) with α=1 and $E\alpha=T0KB$ where KB is the Boltzmann constant. This activation energy does not correspond to any known acceptor energy levels due to Mn doping in Ge, consistent with results shown in [156].

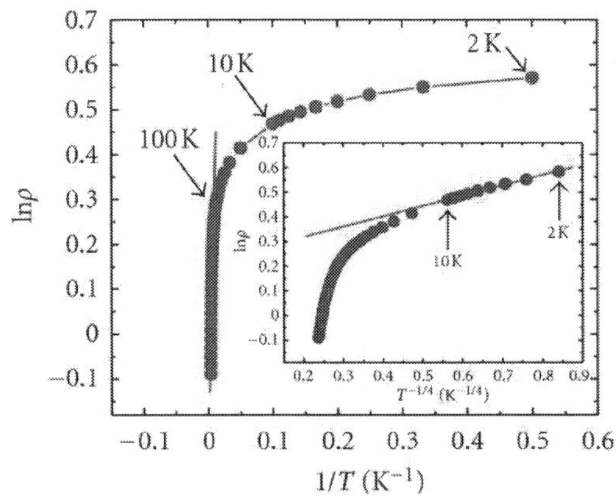

(a)

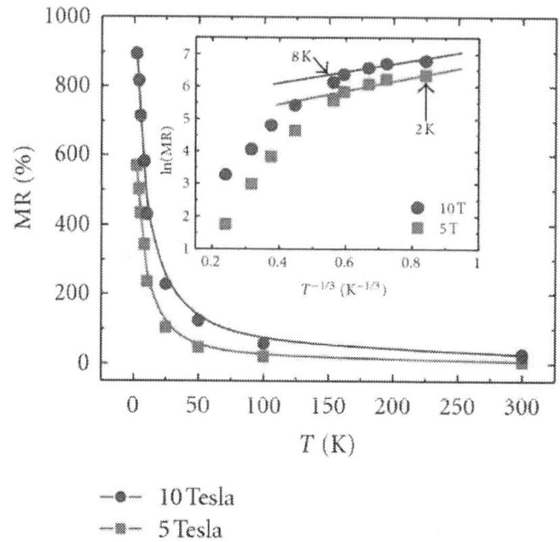

(b)

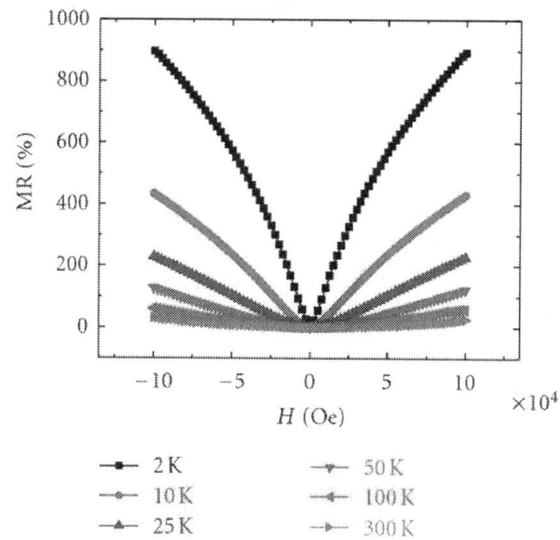

(c)

Figure 14. Magnetotransport measurements for the MnGe nanodot arrays. (a) The temperature-dependent resistivity (lnρ versus T−1) and the inset displays the plot of lnρversus T−1/4. (b) Temperature-dependent MR under fixed magnetic fields of 5 and 10 Tesla and the inset showing the plot of ln(MR) versus T−1/3. (c) Positive MRs at different temperatures and different magnetic fields.

To explore practical applications for our extraordinary nanodot arrays, the MR measurements were performed from 2 to 300 K with an external magnetic field up to 10 Tesla. Figure 14(b) shows the plots of temperature-dependent MR at given magnetic fields (5 and 10 Tesla) for the nanodot arrays. Under a strong magnetic field, the MR in the region of variable range-hopping conduction can be described by [157, 158]

$$\mathrm{MR}(H) = \exp\left[\frac{c}{(\lambda^2 T)^{1/3}}\right] - 1 , \tag{3}$$

where the magnetic length λ equals $(c\hbar/eH)1/2$ and C is a field- and temperature-independent constant. Note that (3) is only valid in a strong-field limit [157–159]. The inset in Figure 14(b) shows the best fitting results, in which a linear behavior of MR versus $T-1/3$ is obtained, further confirming the hopping conduction mechanisms ($T\leq 8$ K). Note that the absolute values of MRs were used for the fitting purpose. These fitting results are reasonably close to the obtained hopping regions determined from the zero-magnetic-field resistivity measurements ($T\leq 10$K, Figure 14(a)).

It is striking to observe that the coherent MnGe nanodot arrays present a large and positive MR up to 900% at 2 K (Figure 14(c)). Traditionally, the positive MR is attributed to the Lorentz force in the semiconductor matrix, which deflects the carriers during the transport process [159]. The resulting MR is positive and proportional to $(\mu H)2$ under low magnetic fields ($H\leq 1$ Tesla in our case) where μ is the semiconductor mobility (units $m^2V^{-1}S^{-1}$ or T^{-1}) and H is the magnetic field. However, with a simple calculation, the estimated orbital MR is too small to explain the large MR observed from the nanodot arrays. Instead, we anticipate that, besides the effect of orbital MR, the high-density magnetic nanodots could significantly contribute to the large MR ratios due to an enhanced geometric MR effect, from which the current path may be significantly deflected when external magnetic fields were applied to the magnetic nanostructures [57, 160]. To elucidate the underlying physics of the geometrical effect, we consider a thin Hall bar geometry with a measurement current applied in the x-direction, a Hall voltage in the y direction, z direction normal to the sample surface, and an external magnetic field H parallel to z. For semiconductors, the current density and the total electric field can be described by $j=\sigma E$, where the magnetoconductivity tensor is given by [57, 160]

$$(\overline{\sigma}H)=\begin{pmatrix}\frac{\sigma}{1+\beta^2} & \frac{\sigma\beta^2}{1+\beta^2} & 0\\ \frac{-\sigma\beta^2}{1+\beta^2} & \frac{\sigma}{1+\beta^2} & 0\\ 0 & 0 & \sigma\end{pmatrix}. \tag{4}$$

Here, $\beta=\mu H$. At zero magnetic field, β vanishes. The conductivity tensor is diagonal when lacking of the magnetic field; and the current density can be simply described by $j=\sigma E$. Since the electric field is normal to the surface of a metallic inclusion and $j\parallel\sigma E$, the current flowing through the material is concentrated into the metallic region which behaves like a "short circuit" [35]. As a result, the inclusion of metallic clusters can lead to a higher conduction than that of a homogeneous semiconductor [57, 160]. However, at high magnetic fields (β>>1), the off-diagonal terms of (H) dominate. Equivalently, the Hall angle between j and E approaches 90°(jE); and the current becomes tangent to the nanodots. This further indicates that the current is deflected to flow around the nanodots, resembling an "open circuit" state [35]. The transition from the "short circuit" at the zero field to the "open circuit" at high fields produces an increase of resistance, that is, a positive geometrically enhanced MR [160].

Discussions

The introduction of high percentage Mn (x ~ 20%) generates metallic behavior in Mn-doped Ge. Such heavily doped samples possess nanoscale MnGe dots. With a superlattice approach, we have successfully fabricated extraordinarily coherent and self-organized MnGe nanodot arrays embedded in the Ge and GaAs matrixes by low temperature MBE. A high yield of such aligned nanodot arrays was confirmed on different substrates, showing an ideal controllability and reproducibility. More importantly, giant positive magnetoresistances were obtained due to the geometrically enhanced effect. We anticipate that these studies will advance the development of MnGe magnetic semiconductors and/or other similar systems. The obtained coherent and self-assembled nanostructures could be potentially used as the building blocks in the high-density magnetic memories, sensors, MTJs, and other spintronic devices, enabling a new generation of low dissipation magnetoelectronic devices.

6. Summary and Prospective

This paper is a review of theoretical and experimental progress that has been achieved in understanding ferromagnetism and related electronic properties in the Mn_xGe_{1-x} DMSs, and MnGe metallic nanodots. Interest in DMS ferromagnetism is motivated by the possibility to engineer systems that combine many of the technologically useful features of ferromagnetic and semiconducting materials. This goal has been achieved to an impressive degree in (III, Mn)V DMSs, and further progress can be anticipated in the future. However, due to the low Tc of (III, Mn)V DMSs, the spintronics research seemly reaches a critical bottleneck, where achieving a high Tc DMS becomes an intriguing and challenging task. Fortunately, the Mn_xGe_{1-x} material system offers a possible route towards higher Tc. The structural and magnetic properties can be adjusted simply by modifying the system dimensions from 3-d thin films to 0D quantum dots. A high Tc in excess of 400 K can be obtained and is presumably attributed to the quantum confinement effect, which strengthens the hybridation between the localized Mn impurities and iterant holes. Bound magnetic polarons may also exist since this system falls into a regime where Mn concentrations are much larger than that of holes. Nevertheless, the high Tc in the low-dimensional system suggests that the quantum structure exibits extraordinary properties that significantly differ from these of bulk films. The future progress would rely on the precise theoretical understanding of the quantum confinement effect on ferromagnetism.

It is well known that in order to achieve functioning spintronic devices working at ambient temperatures, it requires the following criteria: (i) the ferromagnetic transition temperature should safely exceed room temperature, (ii) the mobile charge carriers should respond strongly to changes in the ordered magnetic state, and (iii) the material should retain fundamental semiconductor characteristics, including sensitivity to doping and light and electric fields produced by gate charges. For more than a decade, these three key issues have been the focus of intense experimental and theoretical research. Progress has been also made in achieving field-controlled ferromagnetism in (III, Mn)As system, even though the controllability remains at low temperatures because of low Tc. Therefore, the critical challenge now is either to continue increasing Tc in (III, Mn)As, or to look for a new DMS system with both high Curie temperature (Tc>>300 K) and the field-controlled ferromagnetism to satisfy all these three criteria. The experimental results of field-controlled ferromagnetism in the $Mn_{0.05}Ge_{0.95}$ QDs suggest that the ferromagnetism in this system sensitively responds to the electrical field via the hole-mediated effect, similar to that in (III, Mn)As system. Therefore, with a much higher Tccompared with III-V DMS, the Mn_xGe_{1-x} nanostructures could become one of the most promising candidates to achieve room-temperature operation.

On the other hand, the metallic MnGe nanodots with a high percentage of Mn (~20%) show perfect lattice coherence with the Ge matrix. Such a ferrromagnet/semiconductor hybrid system is extremely useful owing to their high-quality defect-free interface, where a high efficient spin injection can be realized without a need of oxides such as MgO (spin filter). We also hope to encourage more research efforts in this direction because of the excellent compatibility of MnGe nanodots with the current CMOS technology.

ACKNOWLEDGMENTS

F. Xiu would like to acknowledge startup funds from the Department of Electrical and Computer Engineering, Colleage of Engineering, and Microelectronics Research Center (MRC) at Iowa State University. F. Xiu thanks Professor Kang L. Wang (UCLA, USA), Professor Jin Zou (Queensland University, Australia), and Professor Yong Wang (Zhejiang University, China) for their contributions to this work.

REFERENCES

1. J. S. Kulkarni, O. Kazakova, D. Erts, M. A. Morris, M. T. Shaw, and J. D. Holmes, "Structural and magnetic characterization of $Ge_{0.99}Mn_{0.01}$ nanowire arrays," Chemistry of Materials, vol. 17, no. 14, pp. 3615–3619, 2005.
2. H. Ohno, D. Chiba, F. Matsukura et al., "Electric-field control of ferromagnetism," Nature, vol. 408, no. 6815, pp. 944–946, 2000.
3. K. Brunner, "Si/Ge nanostructures," Reports on Progress in Physics, vol. 65, no. 1, pp. 27–72, 2002.
4. T. C. Schulthess and W. H. Butler, "Electronic structure and magnetic interactions in Mn doped semiconductors," Journal of Applied Physics, vol. 89, no. 11, pp. 7021–7023, 2001.
5. M. van Schilfgaarde and O. N. Mryasov, "Anomalous exchange interactions in III-V dilute magnetic semiconductors," Physical Review B, vol. 63, no. 23, Article ID 233205, 2001.
6. Y. F. Chen, W. N. Lee, J. H. Huang et al., "Growth and magnetic properties of self-assembled (In, Mn)As quantum dots," Journal of Vacuum Science and Technology B, vol. 23, no. 4, pp. 1376–1378, 2005.
7. E. Arras, F. Lançon, I. Slipukhina, et al., "Interface-driven phase separation in multifunctional materials: The case of the ferromagnetic semiconductor GeMn," Physical Review B, vol. 85, no. 11, 10 pages, 2012.

8. C. Liu, F. Yun, and H. Morkoç, "Ferromagnetism of ZnO and GaN: a review," Journal of Materials Science, vol. 16, no. 9, pp. 555–597, 2005.
9. M. Bolduc, C. Awo-Affouda, A. Stollenwerk et al., "Above room temperature ferromagnetism in Mn-ion implanted Si," Physical Review B, vol. 71, no. 3, Article ID 033302, 4 pages, 2005.
10. T. Jungwirth, J. Sinova, J. Masek, J. Kucera, and A. H. MacDonald, "Theory of ferromagnetic (III,Mn)V semiconductors," Reviews of Modern Physics, vol. 78, no. 3, pp. 809–864, 2006.
11. W. Lu and C. M. Lieber, "Semiconductor nanowires," Journal of Physics D, vol. 39, no. 21, article R387, 2006.
12. P. Lyu and K. Moon, "Ferromagnetism in diluted magnetic semiconductor quantum dot arrays embedded in semiconductors," European Physical Journal B, vol. 36, no. 4, pp. 593–598, 2003.
13. M. I. van der Meulen, N. Petkov, M. A. Morris et al., "Single crystalline Ge1−xMnx nanowires as building blocks for nanoelectronics," Nano Letters, vol. 9, no. 1, pp. 50–56, 2009.
14. S. Maekawa, Concepts in Spin Electronics, Oxford University Press, 2006.
15. K. L. Wang, S. G. Thomas, and M. O. Tanner, "SiGe band engineering for MOS, CMOS and quantum effect devices," Journal of Materials Science, vol. 6, no. 5, pp. 311–324, 1995.
16. R. N. Bhatt, M. Berciu, M. P. Kennett, and X. Wan, "Diluted magnetic semiconductors in the low carrier density regime," Journal of Superconductivity: Incorporating Novel Magnetism, vol. 15, no. 1, pp. 71–83, 2002.
17. Y. J. Cho, C. H. Kim, H. S. Kim et al., "Ferromagnetic Ge1−xMx (M = Mn, Fe, and Co) nanowires,"Chemistry of Materials, vol. 20, no. 14, pp. 4694–4702, 2008.
18. M. Berciu and R. N. Bhatt, "Effects of disorder on ferromagnetism in diluted magnetic semiconductors,"Physical Review Letters, vol. 87, no. 10, Article ID 107203, 4 pages, 2001.
19. L. J. Lauhon, M. S. Gudlksen, D. Wang, and C. M. Lieber, "Epitaxial core-shell and core-multishell nanowire heterostructures," Nature, vol. 420, no. 6911, pp. 57–61, 2002.
20. N. Nepal, M. O. Luen, J. M. Zavada, S. M. Bedair, P. Frajtag, and N. A. El-Masry, "Electric field control of room temperature ferromagnetism in III-N dilute magnetic semiconductor films," Applied Physics Letters, vol. 94, no. 13, Article ID 132505, 3 pages, 2009.
21. T. Kanki, H. Tanaka, and T. Kawai, "Electric control of room temperature ferromagnetism in a Pb $(Zr_{0.2}Ti_{0.8})O_3/La_{0.85}Ba_{0.15}MnO_3$ field-effect transistor," Applied Physics Letters, vol. 89, no. 24, Article ID 242506, 3 pages, 2006.

22. J. Philip, A. Punnoose, B. I. Kim et al., "Carrier-controlled ferromagnetism in transparent oxide semiconductors," Nature Materials, vol. 5, no. 4, pp. 298–304, 2006.
23. D. Chiba, F. Matsukura, and H. Ohno, "Electric-field control of ferromagnetism in (Ga,Mn)As," Applied Physics Letters, vol. 89, no. 16, Article ID 162505, 3 pages, 2006.
24. Y. Rheem, B.-Y. Yoo, W. P. Beyermann, and N. V. Myung, "Magneto-transport studies of single ferromagnetic nanowire," Physica Status Solidi A, vol. 204, no. 12, pp. 4004–4008, 2007.
25. H.-J. Choi, H. K. Seong, J. Chang et al., "Single-crystalline diluted magnetic semiconductor GaN:Mn nanowires," Advanced Materials, vol. 17, no. 11, pp. 1351–1356, 2005.
26. T. Miyoshi, T. Matsui, H. Tsuda, H. Mabuchi, and K. Morii, "Magnetic and electric properties of Mn_5Ge_3/Ge nanostructured films," Journal of Applied Physics, vol. 85, no. 8, Article ID 5372, 3 pages, 1999.
27. Y. D. Park, A. T. Hanbicki, S. C. Erwin et al., "A group-IV ferromagnetic semiconductor: MnxGe1−x,"Science, vol. 295, no. 5555, pp. 651–654, 2002.
28. F. Tsui, L. He, L. Ma et al., "Novel germanium-based magnetic semiconductors," Physical Review Letters, vol. 91, no. 17, Article ID 177203, 4 pages, 2003.
29. F. D'Orazio, F. Lucari, N. Pinto, L. Morresi, and R. Murri, "Toward room temperature ferromagnetism of Ge:Mn systems," Journal of Magnetism and Magnetic Materials, vol. 272–276, part 3, pp. 2006–2007, 2004.
30. O. Kazakova, J. S. Kulkarni, J. D. Holmes, and S. O. Demokritov, "Room-temperature ferromagnetism in Ge1−xMnx nanowires," Physical Review B, vol. 72, no. 9, Article ID 094415, 6 pages, 2005.
31. P. Li, J. F. Wendelken, J. Shen, L. C. Feldman, J. R. Thompson, and H. H. Weitering, "Magnetism inMnxGe1−x semiconductors mediated by impurity band carriers," Physical Review B, vol. 72, no. 19, Article ID 195205, 9 pages, 2005.
32. E. Demidov, Y. A. Danilov, V. V. Podol'skiĭ, V. P. Lesnikov, M. V. Sapozhnikov, and A. I. Suchkov, "Ferromagnetism in epitaxial germanium and silicon layers supersaturated with manganese and iron impurities," JETP Letters, vol. 83, no. 12, pp. 568–571, 2006.
33. M. Jamet, A. Barski, T. Devillers et al., "High-curie-temperature ferromagnetism in self-organizedGe1−xMnx nanocolumns," Nature Materials, vol. 5, no. 8, pp. 653–659, 2006.
34. B. A. Collins, Y. S. Chu, L. He, Y. Zhong, and F. Tsui, "Dopant stability and strain states in Co and Mn-doped Ge (001) epitaxial films," Physical Review B, vol. 77, no. 19, Article ID 193301, 4 pages, 2008.

35. M. Ogawa, X. Han, Z. Zhao, Y. Wang, K. L. Wang, and J. Zou, "Mn distribution behaviors and magnetic properties of GeMn films grown on Si (001) substrates," Journal of Crystal Growth, vol. 311, no. 7, pp. 2147–2150, 2009.
36. R. Tsuchida, J. T. Asubar, Y. Jinbo, and N. Uchitomi, "MBE growth and properties of GeMn thin films on (001) GaAs," Journal of Crystal Growth, vol. 311, no. 3, pp. 937–940, 2009.
37. S. Cho, S. Choi, S. C. Hong et al., "Ferromagnetism in Mn-doped Ge," Physical Review B, vol. 66, no. 3, Article ID 033303, 3 pages, 2002.
38. S. Pramanik, S. Bandyopadhyay, and M. Cahay, "Spin transport in nanowires," in Proceedings of the 3rd IEEE Nanotechnology Conference (IEEE-NANO '03), vol. 2, pp. 87–90, 2003.
39. J. Chen, K. L. Wang, and K. Galatsis, "Electrical field control magnetic phase transition in nanostructured MnxGe1−x," Applied Physics Letters, vol. 90, no. 1, Article ID 012501, 3 pages, 2007.
40. F. Xiu, Y. Wang, J. Kim, et al., "Electric-field-controlled ferromagnetism in high-Curie-temperature $Mn_{0.05}Ge_{0.95}$ quantum dots," Nature Materials, vol. 9, pp. 337–344, 2010.
41. S. Maekawa, "Spin-dependent transport in magnetic nanostructures," Journal of Magnetism and Magnetic Materials, vol. 272–276, no. 1, pp. E1459–E1463, 2004.
42. X. Lou, C. Adelmann, S. A. Crooker et al., "Electrical detection of spin transport in lateral ferromagnet-semiconductor devices," Nature Physics, vol. 3, no. 3, pp. 197–202, 2007.
43. P. Li, C. Zeng, K. van Benthem et al., "Dopant segregation and giant magnetoresistance in manganese-doped germanium," Physical Review B, vol. 75, no. 20, Article ID 201201, 4 pages, 2007.
44. S. Majumdar, S. Mandal, A. K. Das, and S. K. Ray, "Synthesis and temperature dependent photoluminescence properties of Mn doped Ge nanowires," Journal of Applied Physics, vol. 105, no. 2, Article ID 024302, 5 pages, 2009.
45. H.-K. Seong, U. Kim, E. K. Jeon et al., "Magnetic and electrical properties of single-crystalline mn-doped Ge nanowires," The Journal of Physical Chemistry C, vol. 113, no. 25, pp. 10847–10852, 2009.
46. G. Zhang, K. Tateno, T. Sogawa, and H. Nakano, "Vertically aligned GaP/GaAs core-multishell nanowires epitaxially grown on Si substrate," Applied Physics Express, vol. 1, Article ID 064003, 3 pages, 2008.
47. H.-Y. Wang and M. C. Qian, "Electronic and magnetic properties of Mn/Ge digital ferromagnetic heterostructures: an ab initio investigation," Journal of Applied Physics, vol. 99, no. 8, Article ID 08D705, 3 pages, 2006.

48. Y. Q. Chang, D. B. Wang, X. H. Luo et al., "Synthesis, optical, and magnetic properties of diluted magnetic semiconductor Zn1−xMnxO nanowires via vapor phase growth," Applied Physics Letters, vol. 83, no. 19, pp. 4020–4022, 2003.
49. J. B. Cui and U. J. Gibson, "Electrodeposition and room temperature ferromagnetic anisotropy of Co and Ni-doped ZnO nanowire arrays," Applied Physics Letters, vol. 87, no. 13, Article ID 133108, 3 pages, 2005.
50. J. M. Baik and J. L. Lee, "Fabrication of vertically well-aligned (Zn,Mn)O nanorods with room temperature ferromagnetism," Advanced Materials, vol. 17, no. 22, pp. 2745–2748, 2005.
51. F. J. Brieler, P. Grundmann, M. Fröba et al., "Formation of Zn1−xMnxS nanowires within mesoporous silica of different pore sizes," Journal of the American Chemical Society, vol. 126, no. 3, pp. 797–807, 2004.
52. P. V. Radovanovic, C. J. Barrelet, S. Gradecak, F. Qian, and C. M. Lieber, "General synthesis of manganese-doped II-VI and III-V semiconductor nanowires," Nano Letters, vol. 5, no. 7, pp. 1407–1411, 2005.
53. M. Rovezzi, T. Devillers, E. Arras et al., "Atomic structure of Mn-rich nanocolumns probed by X-ray absorption spectroscopy," Applied Physics Letters, vol. 92, no. 24, 3 pages, 2008.
54. H. C. Jeon, K. J. Chung, K. J. Chung, T. W. Kang, and T. W. Kim, "Enhancement of the ferromagnetic transition temperature in self-assembled (Ga1−xMnx)As quantum wires," Japanese Journal of Applied Physics, vol. 43, pp. L963–L965, 2004.
55. Y. Ji, A. Hoffmann, J. S. Jiang, and S. D. Bader, "Spin injection, diffusion, and detection in lateral spin-valves," Applied Physics Letters, vol. 85, no. 25, pp. 6218–6220, 2004.
56. Y.-J. Zhao, T. Shishidou, and A. J. Freeman, "Ruderman-Kittel-Kasuya-Yosida-like ferromagnetism inMnxGe1−x," Physical Review Letters, vol. 90, no. 4, Article ID 047204, 4 pages, 2003.
57. S. U. Yuldashev, Y. Shon, Y. H. Kwon et al., "Enhanced positive magnetoresistance effect in GaAs with nanoscale magnetic clusters," Journal of Applied Physics, vol. 90, no. 6, pp. 3004–3006, 2001.
58. Stroppa, S. Picozzi, A. Continenza, and A. J. Freeman, "Electronic structure and ferromagnetism of Mn-doped group-IV semiconductors," Physical Review B, vol. 68, no. 15, Article ID 155203, 9 pages, 2003.
59. Y. Ji, A. Hoffmann, J. S. Jiang, J. E. Pearson, and S. D. Bader, "Non-local spin injection in lateral spin valves," Journal of Physics D, vol. 40, no. 5, article S13, pp. 1280–1284, 2007.
60. H. Akai, "Ferromagnetism and its stability in the diluted magnetic semiconductor (In, Mn)As," Physical Review Letters, vol. 81, no. 14, pp. 3002–3005, 1998.

61. T. Dietl, H. Ohno, and F. Matsukura, "Hole-mediated ferromagnetism in tetrahedrally coordinated semiconductors," Physical Review B, vol. 63, no. 19, Article ID 195205, 21 pages, 2001.
62. S. Picozzi and M. Lezaic, "Ab-initio study of exchange constants and electronic structure in diluted magnetic group-IV semiconductors," New Journal of Physics, vol. 10, Article ID 055017, 2008.
63. X. Luo, S. B. Zhang, and S.-H. Wei, "Theory of Mn supersaturation in Si and Ge," Physical Review B, vol. 70, no. 3, Article ID 033308, 3 pages, 2004.
64. S. C. Erwin and A. G. Petukhov, "Self-compensation in manganese-doped ferromagnetic semiconductors," Physical Review Letters, vol. 89, no. 22, Article ID 227201, 4 pages, 2002.
65. G. A. Prinz, "Magnetoelectronics," Science, vol. 282, no. 5394, pp. 1660–1663, 1998.
66. F. J. Jedema, H. B. Heersche, A. T. Filip, J. J. A. Baselmans, and B. J. van Wees, "Electrical detection of spin precession in a metallic mesoscopic spin valve," Nature, vol. 416, no. 6882, pp. 713–716, 2002
67. H. Chen, W. Zhu, E. Kaxiras, and Z. Zhang, "Optimization of Mn doping in group-IV-based dilute magnetic semiconductors by electronic codopants," Physical Review B, vol. 79, no. 23, Article ID 235202, 13 pages, 2009.
68. S. O. Valenzuela and M. Tinkham, "Spin-polarized tunneling in room-temperature mesoscopic spin valves," Applied Physics Letters, vol. 85, no. 24, pp. 5914–5916, 2004.
69. S. Garzon, I. Zutic, and R. A. Webb, "Temperature-dependent asymmetry of the nonlocal spin-injection resistance: evidence for spin nonconserving interface scattering," Physical Review Letters, vol. 94, no. 17, Article ID 176601, 4 pages, 2005.
70. B. I. Shklovskii and A. L. Efros, Electronics Properties of Doped Semiconductors, Spinger, 1984.
71. N. Poli, M. Urech, V. Korenivski, and D. B. Haviland, "Spin-flip scattering at Al surfaces," Journal of Applied Physics, vol. 99, no. 8, Article ID 08H701, 3 pages, 2006.
72. M. Peressi, A. Debernardi, S. Picozzi, F. Antoniella, and A. Continenza, "Half-metallic Mn-doped Si(x)Ge(1−x) alloys: a first principles study," Computational Materials Science, vol. 33, no. 1–3, pp. 125–131, 2005.
73. T. Kimura and Y. Otani, "Local domain structure of exchange-coupled NiFe/CoO nanowire probed by nonlocal spin valve measurement," Journal of Applied Physics, vol. 103, no. 8, Article ID 083915, 5 pages, 2008.
74. M. Johnson and R. H. Silsbee, "Interfacial charge-spin coupling: injection and detection of spin magnetization in metals," Physical Review Letters, vol. 55, no. 17, pp. 1790–1793, 1985.

75. S. Patibandla, S. Pramanik, S. Bandyopadhyay, and G. C. Tepper, "Spin relaxation in a germanium nanowire," Journal of Applied Physics, vol. 100, no. 4, Article ID 044303, 5 pages, 2006.
76. D. Kelly, J.-E. Wegrowe, T.-K. Truong, X. Hoffer, and J.-P. Ansermet, "Spin-polarized current-induced magnetization reversal in single nanowires," Physical Review B, vol. 68, no. 13, Article ID 134425, 13 pages, 2003.
77. C. Jaeger, C. Bihler, T. Vallaitis et al., "Spin-glass-like behavior of Ge:Mn," Physical Review B, vol. 74, no. 4, Article ID 045330, 10 pages, 2006.
78. S. Ahlers, D. Bougeard, H. Riedl et al., "Ferromagnetic Ge(Mn) nanostructures," Physica E, vol. 32, no. 1-2, pp. 422–425, 2006.
79. S. Ahlers, D. Bougeard, N. Sircar et al., "Magnetic and structural properties of GexMn1−x films: precipitation of intermetallic nanomagnets," Physical Review B, vol. 74, no. 21, Article ID 214411, 8 pages, 2006.
80. C. Bihler, C. Jaeger, T. Vallaitis et al., "Structural and magnetic properties of Mn_5Ge_3 clusters in a dilute magnetic germanium matrix," Applied Physics Letters, vol. 88, no. 11, Article ID 112506, 3 pages, 2006.
81. D. Bougeard, S. Ahlers, A. Trampert, N. Sircar, and G. Abstreiter, "Clustering in a precipitate-free GeMn magnetic semiconductor," Physical Review Letters, vol. 97, no. 23, Article ID 237202, 4 pages, 2006.
82. T. Devillers, M. Jamet, A. Barski et al., "Structure and magnetism of self-organized Ge1−xMnxnanocolumns on Ge(001)," Physical Review B, vol. 76, no. 20, Article ID 205306, 12 pages, 2007.
83. Y. D. Park, A. Wilson, A. T. Hanbicki et al., "Magnetoresistance of Mn:Ge ferromagnetic nanoclusters in a diluted magnetic semiconductor matrix," Applied Physics Letters, vol. 78, no. 18, pp. 2739–2741, 2001.
84. Y. Wang, J. Zou, Z. Zhao, X. Han, X. Zhou, and K. L. Wang, "Direct structural evidences of $Mn_{11}Ge_8$and Mn_5Ge_2 clusters in $Ge_{0.96}Mn_{0.04}$ thin films," Applied Physics Letters, vol. 92, no. 10, Article ID 101913, 3 pages, 2008.
85. N. Pinto, L. Morresi, M. Ficcadenti et al., "Magnetic and electronic transport percolation in epitaxialGe1−xMnx films," Physical Review B, vol. 72, no. 16, Article ID 165203, 7 pages, 2005.
86. E. S. Park, D. H. Kim, and W. T. Kim, "Parameter for glass forming ability of ternary alloy systems,"Applied Physics Letters, vol. 86, no. 6, Article ID 061907, 3 pages, 2005.
87. H.-T. Lin, W.-J. Huang, S.-H. Wang, H.-H. Lin, and T.-S. Chin, "Carrier-mediated ferromagnetism in p-Si(100) by sequential ion-implantation of B and Mn," Journal of Physics, vol. 20, no. 9, Article ID 095004, 2008.

88. Verna, L. Ottaviano, M. Passacantando et al., "Ferromagnetism in ion implanted amorphous and nanocrystalline MnxGe1−x," Physical Review B, vol. 74, no. 8, Article ID 085204, 12 pages, 2006.
89. F. D'Orazio, F. Lucari, M. Passacantando, P. P. S. Santucci, and A. Verna, "Magneto-optical study of Mn ions implanted in Ge," in Proceedings of the IEEE International of Magnetics Conference (INTERMAG Europe '02), 2002.
90. L. Liu, N. Chena, Z. Yina, F. Yanga, J. Zhoua, and F. Zhang, "Investigation of Mn-implanted n-type Ge,"Journal of Crystal Growth, vol. 265, no. 3-4, pp. 466–470, 2004.
91. L. Lifeng, C. Nuofu, C. Chenlong, L. Yanli, Y. Zhigang, and Y. Fei, "Magnetic properties of Mn-implanted n-type Ge," Journal of Crystal Growth, vol. 273, no. 1-2, pp. 106–110, 2004.
92. M. Passacantando, L. Ottaviano, F. D'Orazio et al., "Growth of ferromagnetic nanoparticles in a diluted magnetic semiconductor obtained by Mn^+ implantation on Ge single crystals," Physical Review B, vol. 73, no. 19, Article ID 195207, 5 pages, 2006
93. L. Ottaviano, M. Passacantando, A. Verna et al., "Microscopic investigation of the structural and electronic properties of ion implanted Mn-Ge alloys," Physica Status Solidi A, vol. 204, no. 1, pp. 136–144, 2007.
94. E. Biegger, L. Staheli, M. Fonin, U. Rudiger, and Y. S. Dedkov, "Intrinsic ferromagnetism versus phase segregation in Mn-doped Ge," Journal of Applied Physics, vol. 101, no. 10, Article ID 103912, 5 pages, 2007.
95. S. Majumdar, A. K. Das, and S. K. Ray, "Magnetic semiconducting diode of p-Ge1−xMnx/n-Ge layers on silicon substrate," Applied Physics Letters, vol. 94, no. 12, Article ID 122505, 3 pages, 2009.
96. R. Gunnella, L. Morresi, N. Pinto et al., "Magnetization of epitaxial MnGe alloys on Ge(111) substrates,"Surface Science, vol. 577, no. 1, pp. 22–30, 2005.
97. L. Liu, N. Chen, Y. Wang et al., "Growth and properties of magnetron cosputtering grown MnxGe1−xon Si (001)," Solid State Communications, vol. 137, no. 3, pp. 126–128, 2006.
98. H. Li, Y. Wu, T. Liu, S. Wang, Z. Guo, and T. Osipowicz, "Magnetic and transport properties of Ge:Mn granular system," Thin Solid Films, vol. 505, no. 1-2, pp. 54–56, 2006.
99. W. Zhu, H. H. Weitering, E. G. Wang, E. Kaxiras, and Z. Zhang, "Contrasting growth modes of Mn on Ge(100) and Ge(111) surfaces: subsurface segregation versus intermixing," Physical Review Letters, vol. 93, no. 12, Article ID 126102, 4 pages, 2004.
100. H. Li, Y. Wu, Z. Guo, P. Luo, and S. Wang, "Magnetic and electrical transport properties of Ge1−xMnxthin films," Journal of Applied Physics, vol. 100, no. 10, Article ID 103908, 9 pages, 2006.

101.F. D'Orazio, F. Lucari, S. Santucci et al., "Magneto-optical properties of epitaxial MnxGe1−x films,"Journal of Magnetism and Magnetic Materials, vol. 262, no. 1, pp. 158–161, 2003.

102.C. Zeng, Y. Yao, Q. Niu, and H. H. Weitering, "Linear magnetization dependence of the intrinsic anomalous hall effect," Physical Review Letters, vol. 96, no. 3, Article ID 037204, 4 pages, 2006.

103.H. Liu and P. Reinke, "Formation of manganese nanostructures on the Si(100)−(2×1) surface," Surface Science, vol. 602, no. 4, pp. 986–992, 2008.

104.P. Gambardella, L. Claude, S. Rusponi et al., "Surface characterization of MnxGe1−x and Cr_yMnxGe1−x−y dilute magnetic semiconductors," Physical Review B, vol. 75, no. 12, Article ID 125211, 7 pages, 2007.

105.Y. M. Cho, S. S. Yu, Y. E. Ihm et al., "Neutron irradiation effects on polycrystalline Ge1−xMnx thin films grown by MBE," Current Applied Physics, vol. 6, no. 3, pp. 482–485, 2006.

106.L. Morresi, N. Pinto, M. Ficcadenti, R. Murri, F. D'Orazio, and F. Lucari, "Magnetic and transport polaron percolation in diluted GeMn films," Materials Science and Engineering B, vol. 126, no. 2-3, pp. 197–201, 2006.

107.S. Sugahara, K. L. Lee, S. Yada, and M. Tanaka, "Precipitation of amorphous ferromagnetic semiconductor phase in epitaxially grown Mn-doped Ge thin films," Japanese Journal of Applied Physics, vol. 44, no. 46–49, pp. L1426–L1429, 2005.

108.Y. Wang, J. Zou, Z. Zhao, X. Han, X. Zhou, and K. L. Wang, "Mn behavior in $Ge_{0.96}$ $Mn_{0.04}$ magnetic thin films grown on Si," Journal of Applied Physics, vol. 103, no. 6, Article ID 066104, 3 pages, 2008.

109.M. Passacantando, L. Ottaviano, V. Grossi et al., "Magnetic response of Mn-doped amorphous porous Ge fabricated by ion-implantation," Nuclear Instruments and Methods in Physics Research B, vol. 257, no. 1-2, pp. 365–368, 2007.

110.Y. X. Chen, S. S. Yan, Y. Fang et al., "Magnetic and transport properties of homogeneous MnxGe1−xferromagnetic semiconductor with high Mn concentration," Applied Physics Letters, vol. 90, no. 5, Article ID 052508, 3 pages, 2007.

A. Verna, F. D'Orazio, L. Ottaviano et al., "Magneto-optical investigation of high temperature ion implanted MnxGe1−x alloy: evidence for multiple contributions to the magnetic response," Physica Status Solidi A, vol. 204, no. 1, pp. 145–151, 2007.

111.N. Pinto, L. Morresi, R. Gunnella et al., "Growth and magnetic properties of MnGe films for spintronic application," Journal of Materials Science, vol. 14, no. 5–7, pp. 337–340, 2003.

112.S. S. Yu, T. T. L. Anh, Y. E. Ihm et al., "Magneto-transport properties of amorphous Ge1−xMnx thin films," Current Applied Physics, vol. 6, no. 3, pp. 545–548, 2006.
113.S. Yada, S. Sugahara, and M. Tanaka, "Magneto-optical and magnetotransport properties of amorphous ferromagnetic semiconductor Ge1−xMnx thin films," Applied Physics Letters, vol. 93, no. 19, Article ID 193108, 3 pages, 2008.
114.P. Gambardella, H. Brune, S. S. Dhesi et al., "Paramagnetic Mn impurities on Ge and GaAs surfaces,"Physical Review B, vol. 72, no. 4, Article ID 045337, 5 pages, 2005.
115.F. Tsui, B. A. Collins, L. He et al., "Combinatorial synthesis and characterization of a ternary epitaxial film of Co and Mn doped Ge (001)," Applied Surface Science, vol. 254, no. 3, pp. 709–713, 2007.
116.Paul and B. Sanyal, "Chemical and magnetic interactions in Mn-and Fe-codoped Ge diluted magnetic semiconductors," Physical Review B, vol. 79, no. 21, Article ID 214438, 5 pages, 2009.
117.R. R. Gareev, Y. V. Bugoslavsky, R. Schreiber, A. Paul, M. Sperl, and M. Döppe, "Carrier-induced ferromagnetism in Ge(Mn,Fe) magnetic semiconductor thin-film structures," Applied Physics Letters, vol. 88, no. 22, Article ID 222508, 3 pages, 2006.
118.T. Fukushima, K. Sato, H. Katayama-Yoshida, and P. H. Dederichs, "Spinodal decomposition under layer by layer growth condition and high curie temperature quasi-one-dimensional nano-structure in dilute magnetic semiconductors," Japanese Journal of Applied Physics, vol. 45, no. 12–16, pp. L416–L418, 2006.
119.M. Holub, S. Chakrabarti, S. Fathpour, P. Bhattacharya, Y. Lei, and S. Ghosh, "Mn-doped InAs self-organized diluted magnetic quantum-dot layers with Curie temperatures above 300 K," Applied Physics Letters, vol. 85, no. 6, pp. 973–975, 2004.
120.Y. H. Zheng, J. H. Zhao, J. F. Bi et al., "Cr-Doped InAs self-organized diluted magnetic quantum dots with room-temperature ferromagnetism," Chinese Physics Letters, vol. 24, no. 7, article 092, pp. 2118–2121, 2007.
121.Arrott, "Criterion for ferromagnetism from observations of magnetic isotherms," Physical Review, vol. 108, no. 6, pp. 1394–1396, 1957.
122.S. M. Sze, Physics of Semiconductor Devices, John Wiley & Sons, New York, NY, USA, 3rd edition, 2007.
I. V. Ovchinnikov and K. L. Wang, "Voltage sensitivity of Curie temperature in ultrathin metallic films,"Physical Review B, vol. 80, no. 1, Article ID 012405, 4 pages, 2009.

123. T. Dietl and J. Spalek, "Effect of thermodynamic fluctuations of magnetization on the bound magnetic polaron in dilute magnetic semiconductors," Physical Review B, vol. 28, no. 3, pp. 1548–1563, 1983.
124. D. Kitchen, A. Richardella, P. Roushan, J.-M. Tang, M. E. Flatt, and A. Yazdani, "Hole-mediated interactions of Mn acceptors on GaAs (110) (invited)," Journal of Applied Physics, vol. 101, no. 9, Article ID 09G515, 6 pages, 2007.
125. S. W. Lim, M. C. Jeong, M. H. Ham, and J. M. Myoung, "Hole-mediated ferromagnetic properties inZn1−xMnxO Thin Films," Japanese Journal of Applied Physics, vol. 43, no. 2B, pp. L280–L283, 2004.
126. H. C. Jeon, Y. S. Jeong, T. W. Kang et al., "(In1−xMnx)As diluted magnetic semiconductor quantum dots with above room temperature ferromagnetic transition," Advanced Materials, vol. 14, no. 23, pp. 1725–1728, 2002.
127. D. D. Awschalom, D. Loss, and N. Samarth, Eds., Semiconductor Spintronics and Quantum Computation, Springer, New York, NY, USA, 2002.
128. D. Chiba, F. Matsukura, and H. Ohno, "Electrical magnetization reversal in ferromagnetic III-V semiconductors," Journal of Physics D, vol. 39, no. 13, article R01, pp. R215–R225, 2006.
129. S. Datta and B. Das, "Electronic analog of the electro-optic modulator," Applied Physics Letters, vol. 56, no. 7, pp. 665–667, 1990. · ·
130. Appelbaum and D. J. Monsma, "Transit-time spin field-effect transistor," Applied Physics Letters, vol. 90, no. 26, Article ID 262501, 2007.
131. M. Nazmul, S. Kobayashi, S. Sugahara, and M. Tanaka, "Electrical and optical control of ferromagnetism in III-V semiconductor heterostructures at high temperature (~100 K)," Japanese Journal of Applied Physics, vol. 43, no. 2A, pp. L233–L236, 2004.
132. H. C. Koo, J. H. Kwon, J. Eom, J. Chang, S. H. Han, and M. Johnson, "Control of spin precession in a spin-injected field effect transistor," Science, vol. 325, no. 5947, pp. 1515–1518, 2009.
133. D. Chiba, M. Sawicki, Y. Nishitani, Y. Nakatani, F. Matsukura, and H. Ohno, "Magnetization vector manipulation by electric fields," Nature, vol. 455, no. 7212, pp. 515–518, 2008.
134. H. Boukari, P. Kossacki, M. Bertolini et al., "Light and electric field control of ferromagnetism in magnetic quantum structures," Physical Review Letters, vol. 88, no. 20, Article ID 207204, 4 pages, 2002.
135. T. Dietl, H. Ohno, F. Matsukura, J. Cibert, and D. Ferrand, "Zener model description of ferromagnetism in zinc-blende magnetic semiconductors," Science, vol. 287, no. 5455, pp. 1019–1022, 2000. · ·
136. F. Matsukura, H. Ohno, A. Shen, and Y. Sugawara, "Transport properties and origin of ferromagnetism in (Ga,Mn)As," Physical Review B, vol. 57, no. 4, pp. R2037–R2040, 1998.

137. M. Yagi, K.-I. Noba, and Y. Kayanuma, "Self-consistent theory for ferromagnetism induced by photo-excited carriers," Journal of Luminescence, vol. 94-95, pp. 523–527, 2001.
138. T. Dietl and H. Ohno, "Engineering magnetism in semiconductors," Materials Today, vol. 9, no. 11, pp. 18–26, 2006.
139. M. Weisheit, S. Fähler, A. Marty, Y. Souche, C. Poinsignon, and D. Givord, "Electric field-induced modification of magnetism in thin-film ferromagnets," Science, vol. 315, no. 5810, pp. 349–351, 2007.
140. M. Tanaka, "Ferromagnet (MnAs)/III-V semiconductor hybrid structures," Semiconductor Science and Technology, vol. 17, no. 4, pp. 327–341, 2002.
141. Y. Wang, F. Xiu, Y. Wang et al., "Coherent magnetic semiconductor nanodot arrays," Nanoscale Research Letters, vol. 6, no. 1, 2011.
142. T. Dietl, "From our readers: self-organized growth controlled by charge states of magnetic impurities,"Nature Materials, vol. 5, no. 9, p. 673, 2006.
143. S. Kuroda, N. Nishizawa, K. Takita et al., "Origin and control of high-temperature ferromagnetism in semiconductors," Nature Materials, vol. 6, no. 6, pp. 440–446, 2007.
144. G. Kioseoglou, A. T. Hanbicki, J. M. Sullivan et al., "Electrical spin injection from an n-type ferromagnetic semiconductor into a III-V device heterostructure," Nature Materials, vol. 3, no. 11, pp. 799–803, 2004.
145. T. Dietl, D. D. Awschalom, M. Kaminska, and H. Ohno, Spintronics, Elsevier, 2008.
146. J. C. Slater, "Atomic radii in crystals," The Journal of Chemical Physics, vol. 41, no. 10, article 3199, 6 pages, 1964.
147. Q. H. Xie, A. Madhukar, P. Chen, and N. P. Kobayashi, "Vertically self-organized InAs quantum box islands on GaAs(100)," Physical Review Letters, vol. 75, no. 13, pp. 2542–2545, 1995.
148. G. S. Solomon, J. A. Trezza, A. F. Marshall, and J. S. Harris, "Vertically aligned and electronically coupled growth induced InAs islands in GaAs," Physical Review Letters, vol. 76, no. 6, pp. 952–955, 1996.
149. X. Z. Liao, J. Zou, D. J. H. Cockayne et al., "Annealing effects on the microstructure of Ge/Si(001) quantum dots," Applied Physics Letters, vol. 79, no. 9, pp. 1258–1260, 2001.
150. S. M. Sze, Physics of Semiconductor Devices, John Wiley & Sons, New York, NY, USA, 2nd edition, 1981.
151. Y. Ohno, D. K. Young, B. Beschoten, F. Matsukura, H. Ohno, and D. D. Awschalom, "Electrical spin injection in a ferromagnetic semiconductor heterostructure," Nature, vol. 402, no. 6763, pp. 790–792, 1999.
152. Y. J. Ma, Z. Zhang, F. Zhou, L. Lu, A. Jin, and C. Gu, "Hopping conduction in single ZnO nanowires,"Nanotechnology, vol. 16, no. 6, pp. 746–749, 2005.

153. J. P. Han, M. R. Shen, W. W. Cao, A. M. R. Senos, and P. Q. Mantas, "Hopping conduction in Mn-doped ZnO," Applied Physics Letters, vol. 82, no. 1, pp. 67–69, 2003.
154. N. Pinto, L. Morresi, M. Ficcadenti et al., "Magnetic and electronic transport percolation in epitaxialGe1−xMnx films," Physical Review B, vol. 72, no. 16, Article ID 165203, 7 pages, 2005.
155. B. I. Schlovskii and A. L. Efros, Electronics Properties of Doped Semiconductors, Spinger, Berlin, Germany, 1984.
156. H. Bottger and V. Bryksin, Hopping Conduction in Solids, Akademie, Berlin, Germany, 1985.
157. K. Ganesan and H. L. Bhat, "Growth, magnetotransport, and magnetic properties of ferromagnetic (In,Mn)Sb crystals," Journal of Applied Physics, vol. 103, no. 4, Article ID 043701, 6 pages, 2008.
158. S. A. Solin, T. Thio, D. R. Hines, and J. J. Heremans, "Enhanced room-temperature geometric magnetoresistance in inhomogeneous narrow-gap semiconductors," Science, vol. 289, no. 5484, pp. 1530–1532, 2000.

CITATION

Faxian Xiu, "Magnetic Mn-Doped Ge Nanostructures," ISRN Condensed Matter Physics, vol. 2012, Article ID 198590, 25 pages, 2012. doi:10.5402/2012/198590

CHAPTER 6

Flexible Circuit Technologies for Biomedical Applications

Damien C. Rodger[1], Wen Li[2], James D. Weiland[1], Mark S. Humayun[1] and Yu-Chong Tai[3]

[1] University of Southern California, Los Angeles, CA, USA
[2] Michigan State University, East Lansing, MI, USA
[3] California Institute of Technology, Pasadena, CA, USA

INTRODUCTION

The human body is an incredibly complex organism that exhibits an impressive resilience to external influences, such as infecting bacteria and inhaled particles. This, however, proves to be a difficult problem to overcome when designing and implementing implantable devices for medical therapy, as the body is supremely primed to resist any artificial intervention. One of the critical parameters to consider is the mechanical compatibility of the implanted device with the tissue of interest. Additionally, the footprint of the device should ideally be as minimal as possible to decrease tissue damage and to minimize encapsulation responses. Until recently, however, it has proven difficult to integrate naturally inflexible solid state circuits with flexible components due to connection complexities and pitch limitations. In addition, material and fabrication limitations have prevented the implementation of thin-film cables, radiofrequency coils, and neural stimulation electrode arrays in implantable devices.

Here, after a brief review of the literature, we discuss novel flexible technologies being developed for biomedical applications, and especially for high density functional neural stimulation and recording. The distinctive parameters of a unique material, namely the semicrystalline thermoplastic parylene C, that make it particularly well-suited as a biocompatible substrate for thin-film biomedical circuits are discussed, explaining the reasoning for its use in a novel paradigm as a substrate and not just as a coating. In addition, we discuss a new packaging scheme that has been developed to enable high lead count interconnects using microfabrication equipment for alignment and patterning, and demonstrate a complete parylene-based stimulation microsystem combining radiofrequency coils with solid state circuits and electrodes in a parylene substrate. This integration of solid state circuits with flexible, biocompatible components gives rejuvenated hope for a new generation of neural prosthetics aimed to provide eyesight to the blind and limb movement to para- and quadriplegic patients through spinal cord stimulation. These technologies can also enable integration of solid state devices with novel bioMEMS sensors in such a manner that devices previously only conceived of are now possible.

OUTER RETINAL DISEASE

Blindness due to such outer retinal diseases as retinitis pigmentosa (RP) and age-related macular degeneration (AMD) affect hundreds of thousands of people worldwide. In fact, it has been estimated by The Eye Diseases Prevalence Research Group and the National Eye Institute that AMD alone will affect three million people in the United States by the year 2020 [1]. In addition, the prevalence of RP has been estimated to be approximately 1 in 4000 [2]. Although the pathogeneses of these retinal photoreceptor diseases are, to date, not entirely understood, it is known that they are largely restricted to the outer retina, and that downstream circuitry, although it undergoes significant remodeling, is relatively spared [3-5]. There exist several possible approaches to thwarting the devastating effects of these diseases. Of these, the surgical, pharmacological, stem cell, and dietary approaches are promising. Laser ablation of leaky blood vessels and diets rich in antioxidants, for example, have been shown to slow the progrcss of AMD, but not to have any effect on its incidence. Effective pharmacologic agents, likewise, have long been elusive. Recent evidence has shown that stem cell therapy could also be a possibility for the treatment of such diseases, by possibly replacing the lost photoreceptors with stem cells capable of maturing into photoreceptors that then make connections with the rest of the retina [6]. However, such treatments, in reality, are still very far away from being used in clinical practice, and have a number of ethical and political barriers to their implementation.

In need of another possible treatment for these profound retinal diseases, in 1994 Humayun *et al.*reported the results of a pioneering study to electrically stimulate vertebrate retinas [7], a bioengineering approach that would possibly bypass the diseased photoreceptors in the visual pathway. In this study, bullfrog eyecups, as well as rabbit eyes (the rabbits were injected intravenously with sodium iodate, a chemical toxic to the retinal pigment epithelium and with other effects on photoreceptors), were stimulated with platinum electrodes. In their report, the following conclusion was made: "Surface electrical stimulation of the inner retina in normal eyes and in eyes with outer retinal degeneration can elicit a localized retinal response." This has since had a profound impact on the field of ophthalmology, by dramatically throwing an engineering solution into the mix of possible treatments for patients with devastating blindness due to outer retinal disease.

As a follow-up to this study, it was shown in 1996 by Humayun *et al.* that this seemingly simple approach of passing a carefully controlled electrical current through an electrode placed directly on the retinal surface to activate the still functional electrically excitable cells of the retina can also elicit visual percepts in humans with otherwise bare or no light-perception vision [8]. Perhaps even more important than the perception itself was the nature of the perception. Subjects did not report seeing streaks or multiple percepts of light simultaneously when stimulating from one electrode, as might be expected if the axons of the ganglion cells, as they made their way to the optic nerve in the nerve fiber layer, were being stimulated. Instead, they reported seeing discrete, highly localized percepts that seemed to correlate very well with the location of stimulation. This suggested that it was not the axons that were the most electrically excitable, but instead the cell bodies of the several retinal ganglion cells (RGCs) or the bipolar, or even possibly amacrine or horizontal cells underlying the electrode. Indeed, it has since been conjectured that it could even be the characteristic right-angle bend in the axon of the RGCs that has maximal electrical excitability [9]. No matter the case, this was the confirmation of the feasibility of an engineering approach to treat human blindness with a "retinal prosthesis."

There are now a large variety of approaches to artificial vision, each with their own advantages and disadvantages. There is this epiretinal approach (Figure 1), in which an electrode array is placed directly on the retina from its anterior aspect. There is the subretinal approach, in which an electrode or photodiode array is placed within the layers of the retina [10], as is also shown in Figure 1 (this can be further subdivided to the *ab interno* and *ab externo* approaches, in which an incision is made within the retina or within the sclera, choriocapillaris, or RPE to insert the device, respectively). There is also the optic nerve approach, in which electrodes are placed around the optic nerve in an attempt to coarsely stimulate the afferent ganglion cell axons [11]. Finally, there are also the cortical approaches, whether they be to the visual cortex [12] or to the lateral geniculate nucleus (LGN) of the thalamus [13, 14], the first stop for information from the retina as it enters the cortex. A complete discussion of the advantages and disadvantages of each of the different approaches is beyond the scope of this text, but can be found in several good review articles on visual and retinal prostheses.

The harbinger of an epiretinal prosthesis was the successful demonstration of a prototype 16-electrode device, fabricated by Second Sight Medical Products, Inc. (Sylmar, CA, USA) in six patients [15]. While clearly not enabling such activities of daily living as newspaper reading and facial recognition, previously completely blind subjects can, for instance, differentiate between a plate, a cup, and a knife, in a high-contrast environment free from background distracters, an undeniably incredible feat of engineering and medicine. Furthermore, subjects have been shown to be able to discriminate direction of movement of parallel white bars on a black background, and can locate white squares within a quadrant of otherwise black space [16]. This implant demonstrated the technology as well as the remarkable ability for the human brain to compensate for low-resolution input. Although this had previously been demonstrated by cochlear prostheses for patients with severe hearing impediments [17], it was unclear whether this plasticity would translate well to visual prostheses. In fact, patients have demonstrated their ability to discriminate large letters simply by *instinctively* scanning the camera mounted on their head back and forth over the image displayed in front of them, a sort of innate edge-detection mechanism, dramatically increasing the capability of the 16-electrode device on their retina. Few scientists disagree, however, that increasing the number of electrodes on the implanted array will dramatically add to the capabilities of the prosthesis as a whole in enabling the patient to better carry out activities of daily living. The prototype device is hand assembled, a fact that limits the resolution possibilities of the device. Indeed, were it still necessary to interconnect integrated circuits by hand, computers would be far less useful to us as they are today. The need exists, then, to bring microelectronics and microfabrication technology to bear on the problem of retinal prosthetic devices. The race is on to build high-density multielectrode arrays in such a way and with such materials that the method is scalable to the needs of long term, high-density retinal stimulation. Recently, a sixty electrode version of the epiretinal prosthesis has been tested and has gained the CE Mark in Europe for sale for the treatment of retinitis pigmentosa. The FDA is now reviewing the results of these same studies for potential approval in the United States as a therapy for outer retinal blindness.

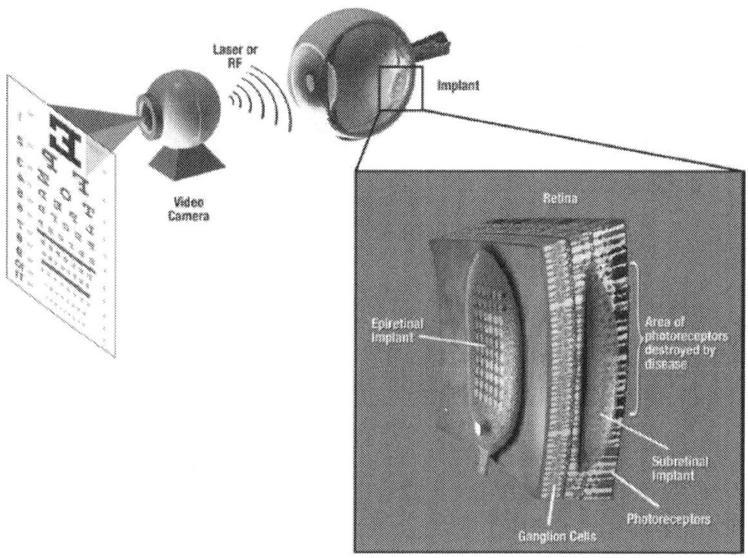

Figure 1. System overview and relative locations of epiretinal and subretinal implants [18].

Spinal Cord Injury

Spinal cord injury (SCI) can occur through a variety of mechanisms. The primary modalities fall into two major categories: trauma (e.g., automobile/motorcycle accident, sports injury such as from diving or horse riding, violence, fall), or disease (e.g., spina bifida or tumor). Spinal cord injuries can affect motor function, sensation, and autonomic functions (e.g., bladder control, breathing). Approximately 50% of SCIs in the United States are classified as complete [19]. This means that the spinal cord has lost the ability to transmit information across a segment within it, cutting off all functional communication from the brain to the nerves below the injury site and resulting in no sensation or voluntary control of motor function below the injury site due to lost input from the brain. This usually results in para- or quadriplegia, depending on the site of injury. Although in most cases the cord is not completely transected or even cut, it is significantly damaged by interruption of blood flow supplying one of its segments or through spinal contusion. Even though the vertebral column protects the cord, when trauma is sufficient to compromise this protective cage, the broken vertebrae can impinge on the cord and crush or destroy the axons within it very quickly, with continued loss of axons over time

[20]. Some estimates pin the prevalence of spinal cord injury in the United States at approximately 250,000 [21], with an incidence of approximately 10,000 to 12,000 per year [22], while others state the prevalence is significantly higher, at around 450,000 [19]. Approximately 55% of spinal cord injuries occur in young victims between 16 and 30 years of age, making it a disease to bear usually for the rest of their lifetime, and more than 80% of victims are males [22]. In some cases of incomplete injury, function can be recovered over time [23]. However, in most cases, some level of impairment is permanent.

The complexity of the spinal cord is undeniable. In fact, despite the popular misconception that the spinal cord serves only as a communication conduit between the brain to the muscles and organs and from the skin back to the brain, it is much more accurate to view the spinal cord as an extension of the brain. While the spinal cord below a complete injury does indeed lose input from the motor cortex, and its ability to send sensations of touch to the brain is completely compromised, the spinal cord is not rendered useless. Even an intact cord does a lot of the primary processing and reflex control without any input to or from the brain.

In patients with spinal cord injury, there are several approaches to rehabilitation. The foot-drop stimulator, wherein stimulation of the peroneal nerve affects localized contraction of ankle dorsiflexors to counteract the problem of foot drag, has been widely studied [24] with mixed results [25-27]. Other functional movements requiring much more coordinated musculature responses are far more difficult with implantable peripheral nerve or muscle stimulators because of the need to control timing as well as pulse amplitudes of likely a large number of electrodes in rapid succession. Skin surface electrodes such as those in the ParaStep system [28, 29] suffer from these problems as well as the problem that many muscle groups are difficult to target from this more remote location. In addition, because in complete SCIs the voluntary input from the motor cortex to initiate such movements is lost, an accessory mechanism for determining the desired motion is warranted. This may require recording electrode arrays in the motor cortex as another component of this system, as well as possible electrical stimulatory feedback (in addition to the visual feedback already present). Any such system, then, is likely to be quite complex and difficult to implement in practice.

We propose a system, which eventually will be completely implantable, that is capable of stimulating the dorsum of the spinal cord in such a manner that modulation of the sensory input to the cord, interneruonal activity within, and even motor output from the cord, is possible. Tonic, subthreshold stimulation, applied at precise times and precise locations along the spinal cord, would likely help sustain or stop locomotor activity with the kind of coordination and rhythmicity already discussed. Perhaps in combination with both locomotor and standing training as well as appropriate pharmacological administration (e.g., quipazine), it is possible that such an array would give those with both complete and incomplete spinal injury the ability to stand and walk once again. In order to provide appropriate proprioceptive input in the case of complete SCI, it would likely be possible to provide an accessory device or muscle stimulator to initiate this type of activity. However, this approach leverages the innate activity and processing power present in the spinal cord to its greatest extent and likely obviates the need for a stimulation control system of great complexity in order to bring about coordinated muscle activity, as is necessary for a peripheral muscle or motor neuron control system. Such electrical stimulation may also, as has been hypothesized in the case of subthreshold retinal stimulation systems as well, have the capability of promoting axon regrowth [30, 31] and facilitating plastic changes in the cord. Such an array would likely need to be conformable to the cord, and would need to be implanted in relatively close apposition to it, either epidurally (from which location the electrical field would need to penetrate several meningeal layers) or subdurally, such that it is closer to the surface of the cord. In the ideal case, the array would be flexible enough to move with the cord during motion and bending such that functional targets are the same in any position. In addition, it should be the case that such an array can also record from the cord so as to recognize returning action potentials from the dorsal root and modulate this input accordingly. This approach, then, requires a high-density array with many electrode sites from which to choose during training and daily activity, as well as the ability to both record from and stimulate the cord, something which current arrays, such as those for pain management, simply cannot do because they are too bulky, inflexible, and of too low a density (up to 16 electrodes).

Introduction to Parylene

Parylene is the trade name for a family of semicrystalline thermoplastic polymers known as the poly(para-xylylenes) (PPX). These were discovered in 1947 by Michael Szwarc in Manchester, England [32]. It was originally deposited in an investigation of aliphatic carbon-hydrogen bonds where the carbon was attached directly to a benzene ring, by heating toluenes and ortho-, meta- and para-xylenes to very high temperatures and looking for degradation products [33]. The very first parylene film deposited, then, was parylene N (with no substitutions on the benzene ring), and was temporarily known as a Szwarcite snakeskin. This deposition process, however, had the disadvantage that gaseous by-products were an inherent part of the method. William Gorham, an employee at Union Carbide, soon thereafter devised an alternate method of parylene deposition that involved the pyrolysis of a dimerized form of the material, di-para-xylylene, or [2.2] paracyclophane. It was largely unknown at the time how to make this dimer in large amounts as it had only been isolated as an impurity in Szwarc's method. However, in 1951 Donald Cram reported a method for making this material in bulk. On February 17, 1965, Union Carbide announced the availability of parylene films and the new vacuum deposition method, known as the Gorham process. There were over 20 types of parylene actually developed, but only three were considered commercially viable: these were parylene N (no chlorines on the benzene ring), parylene C (one chlorine on the benzene ring), and parylene D (with two chlorines on the benzene ring) (Figure 2). However, a new fluorinated version of parylene, parylene HT, has recently become commercially available, and can be deposited in a new parylene deposition system.

In the Gorham vapor-deposition process [34] (depicted in Figure 3), which takes place at vacuum (~25-35 mT, to increase mean free path to the substrate), a charge of parylene dimer is placed in a vaporizer furnace. The dimer evaporates at approximately 130 to 150 °C, and then passes through a very high temperature pyrolysis (~650 to 750 °C) furnace, where the molecule is split into monomers. The monomers enter a chamber that is held at room temperature, and polymerize on all exposed surfaces in the chamber conformally and without pinholes. Residual monomer is collected on a cold trap. Different variants of parylene require varying process conditions, but the method remains essentially the same.

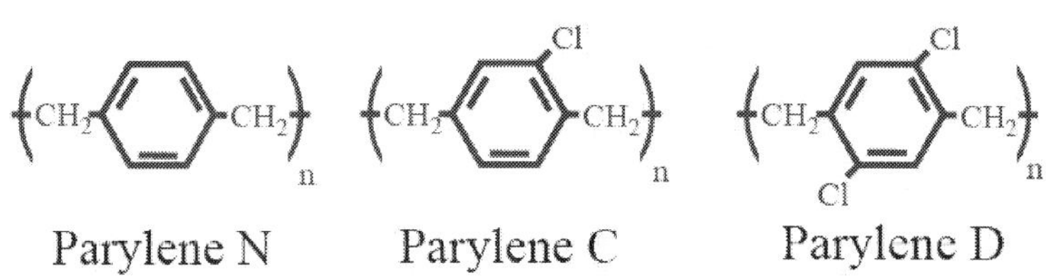

Figure 2. The three originally commercialized parylene variants.

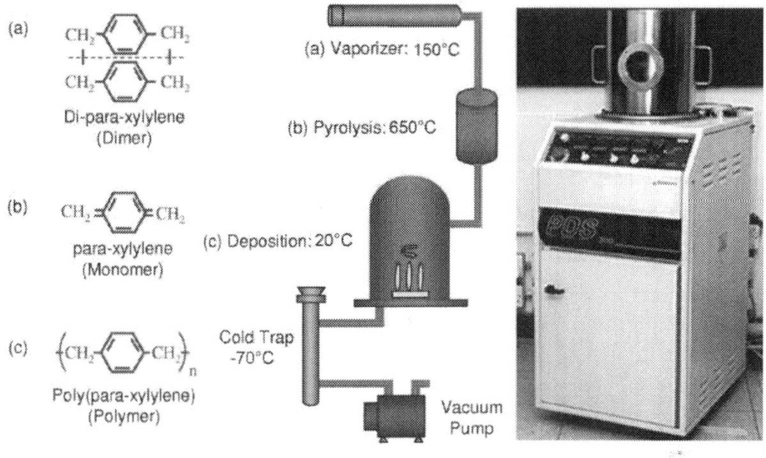

Figure 3. (center) Gorham process for parylene deposition. (left) Chemical structures at corresponding points. (right) PDS 2010 Labcoater System.

The existing variants of parylene have varying properties as well as uses. Parylene N is primarily used as a dielectric and when lubricity and crevice penetration is important. Parylene C is an ISO 10993, United States Pharmacopeia (USP) Class VI material (the highest biocompatibility rating for plastics in the United States) and has excellent water barrier properties. In addition, it has a very large elongation to break. Parylene D has now been largely replaced by parylene HT [35], but is used when mechanical strength is of primary concern. Parylene HT has extremely low coefficients of static and kinetic friction, excellent thermal stability and good water barrier properties, high ultraviolet stability, and is also ISO 10993 biocompatible [35-37]. Finally, very recently, Kishimoto Sangyo Co., Ltd. in Japan has devised additional parylenes in which amino groups have been added to the benzene rings. The amino group may add even more improved biostability, but could also generate bioactivity that may or may not be beneficial to device functionality. We have preliminarily tested these amino parylenes have found them to be compatible with standard parylene processing technology, such as oxygen plasma reactive-ion etching (RIE), in our cleanroom, but they will not be explored further in this text.

The advantages of using parylene, and, more specifically, parylene C, as the structural material for neuroprostheses, when compared with technologies based on the use of other materials such as PDMS, polyimide [38] and silicon [39], include parylene's pinhole-free conformality due to its unique room-temperature chemical vapor deposition process, its low water permeability, its chronic implantability and its high flexibility and mechanical strength (Young's modulus ~4 GPa). The Young's moduli of two other commonly used materials for neuroprostheses, PDMS and polyimide 2611, are graphed alongside that of parylene C in Figure 4. PDMS arrays have been handled by surgeons in our vivarium, and these are often as reported as too floppy and difficult to handle due in part to the low Young's modulus, hence requiring very large thicknesses to handle appropriately. In addition, polyimide 2611 (often chosen because its water permeability is lower than that of other polyimides) has a Young's modulus larger than that of parylene C. It has been suggested that polyimide arrays often are too rigid and can damage the retina. In addition, they tear quite easily. Parylene C handles very well under surgical manipulation, and, as we will show, behaves very well when implanted. In addition, the thickness of parylene films is determined by the mass of dimer placed in the vaporizer. As such, thicknesses can be very thin or very thick, and thicknesses are very repeatable and well controlled, more so than spin-on coatings, especially when deposited over step junctions. Parylene thickness can be controlled so as to match the rigidity to the application of interest, more easily than most other materials. Since parylene is deposited at room temperature (we have verified this using Temp-Plate irreversible temperature recorders traceable to NIST (Wahl Instruments, Inc., Asheville, NC, USA)), the coating process is post-IC compatible. Parylene C is also optically transparent, enabling the anatomy to be seen through the cable and the array during ophthalmic surgery, post-implantation examination, and follow-up. While many groups use parylene C as a coating on their arrays for many of these reasons, we have chosen to use it as the main substrate for our devices [40, 41], a paradigm that leverages these advantages to the greatest extent.

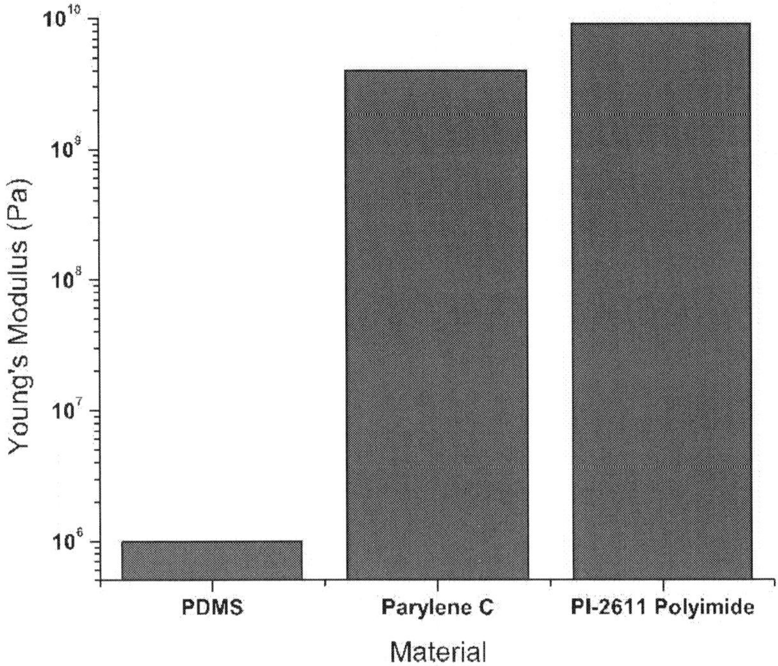

Figure 4. Comparison of Young's moduli of various materials. Ordinate is logarithmic.

Although parylene is known to be biocompatible in many sites of implantation, in order to initially assess the intraocular biocompatibility of the material in the unique immune environment of the eye, an approximately 2 cm × 0.5 cm piece of unmodified 20 μm thick parylene C was implanted in the vitreous cavity of the right eyes of two rabbits for six months. The retinas of the right eyes of both rabbits were compared post-mortem with those of their left eyes that served as controls. Histological evaluation revealed no discernable difference between right and left eyes, indicating that there was no detectable adverse retinitis, choroiditis, endophthalmitis, or scleritis seen as a result of parylene implantation in the vitreous cavity [40]. These results supported the tenet that parylene C is a biocompatible bulk material for an intraocular retinal prosthesis and other ocular implants, and paved the way for the design and fabrication of a flexible electrode arrays and a packaging system using parylene C as the primary substrate. Given these biocompatibility results, we have also been investigating parylene in several other ocular implants, with excellent results to date [42-44]. Similar experiments were performed with parylene C implanted on the spinal cord of mice. The arrays were well tolerated, with no obvious immune reaction or gliosis.

Single-metal-layer Flexible MEAs

Fabrication

Single-metal-layer parylene C-based electrode arrays are fabricated as shown in Figure 5. A photoresist sacrificial layer is optionally spun on a standard silicon wafer. Approximately 8 μm of parylene C is then vapor deposited in a PDS2010 system (Specialty Coating Systems, Indianapolis, IN, USA) on the entire wafer. An LOR3B photoresist layer (Microchem Corp., Newton, MA, USA) and an AZ1518 layer (AZ Electronic Materials, Branchburg, NJ, USA) are spun on top of the parylene, exposed in a 10X reduction GCA Mann 4800 DSW wafer stepper (General Signal Corporation, Stamford, CT, USA) or a Kasper 2001 contact aligner (Kasper Instruments, Inc., Sunnyvale, CA, USA) depending on the required resolution of the electrode array, and developed to achieve a liftoff pattern comprising contacts, conductive traces, and electrodes. After hard bake, approximately 2000 Å to 5000 Å of platinum, with or without a 200 Å titanium layer, is then e-beam evaporated (SE600 RAP, CHA Industries, Fremont, CA, USA) on the wafer. The subsequent photoresist strip generates the desired single-layer metallization pattern. An approximately 7 μm thick coating of parylene C is then deposited, followed by a spin coating of photoresist. This photoresist etch mask is exposed over the areas of the electrodes and contact pads and to pattern the overall array geometry, and the entire wafer is then subjected to an RIE in oxygen plasma, removing the parylene insulation over the electrodes and the parylene surrounding the array. The photoresist mask is then removed with solvent. Finally, if a sacrificial photoresist layer was used, the array is released from the substrate in an acetone bath. If no sacrificial layer was used, it is peeled from the silicon in a water bath. Ultimately, for most cases, the sacrificial photoresist layer is unnecessary, and can often complicate array fabrication due to cracking while under process. The arrays can be easily released from a natively oxidized silicon surface by placing them in a deionized water bath and peeling them from their edge. The water will then release the rest of the structure due to the hydrophobicity of the underlying parylene surface.

A single-layer square-grid electrode array, consisting of 256 Ti/Pt thin-film electrodes 125 μm in diameter in a 16 × 16 grid with connecting lines of 12 μm pitch fabricated in the manner of Figure 5 is shown in Figure 6. An SEM highlighting the typical electrode morphology in such structures is shown in Figure 7. As can be seen, the parylene covering the electrode has been completely removed, whereas the incoming trace remains conformally coated with the material.

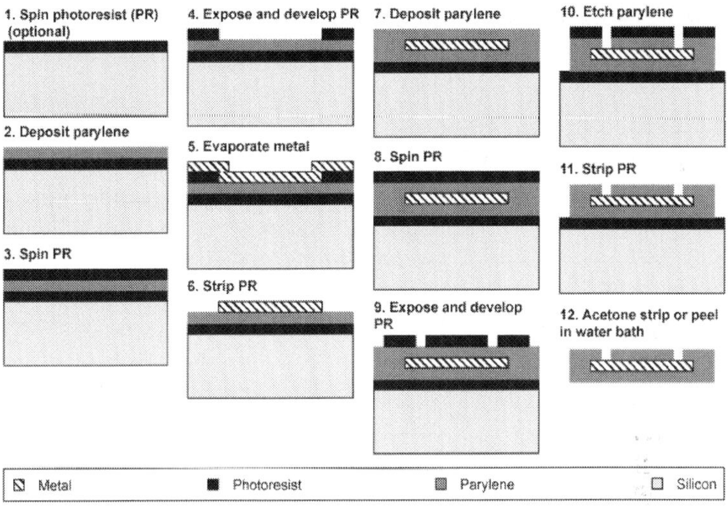

Figure 5. Fabrication process for parylene-based single-metal-layer flexible MEAs.

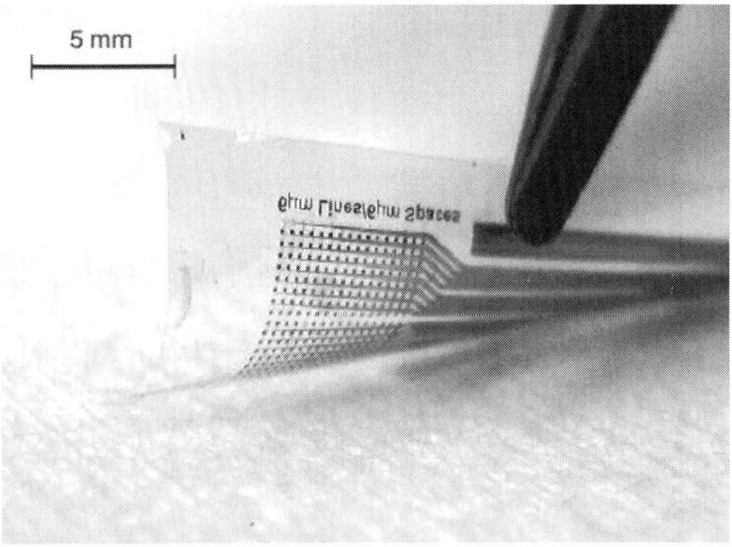

Figure 6. Photograph of Ti/Pt electrode array of 256 electrodes and lines of 12 μm pitch.

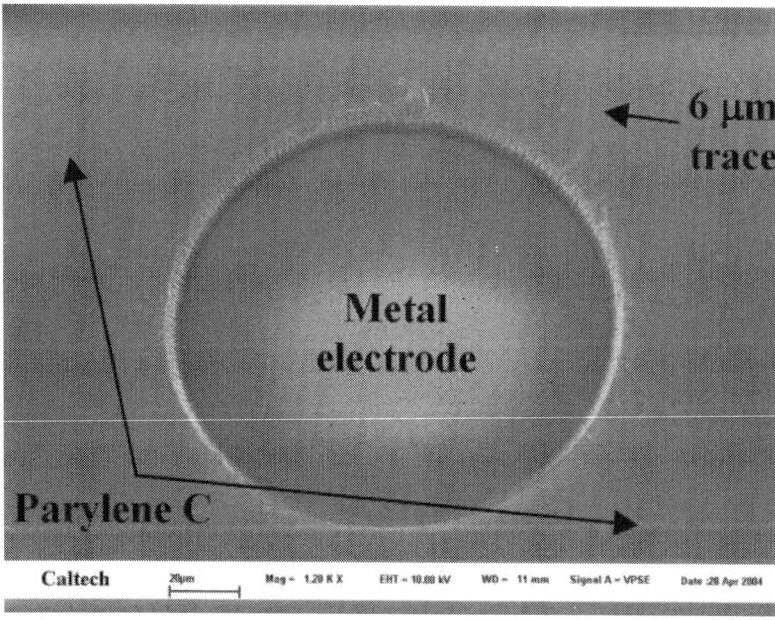

Figure 7. SEM of electrode morphology showing parylene C insulation surrounding exposed metal electrode.

In Vitro Retinal Recording and Stimulation

Parylene C-based arrays of thin-film platinum electrodes, comprising four 200 µm diameter stimulating electrodes and 56 recording electrodes of 10 µm diameter were fabricated according to the single-metal-layer process on a glass substrate. These were placed in a bicarbonate perfusate under a microscope and connected to a stimulus generator and preamplification board (Multi Channel Systems MCS GmbH, Reutlingen, Germany) [45]. As shown in Figure 8, a retina isolated from larval tiger salamander (*Ambystoma tigrinum*) was placed RGC side down on the array (to simulate epiretinal stimulation), and a remote platinum ground electrode was introduced to the bath.

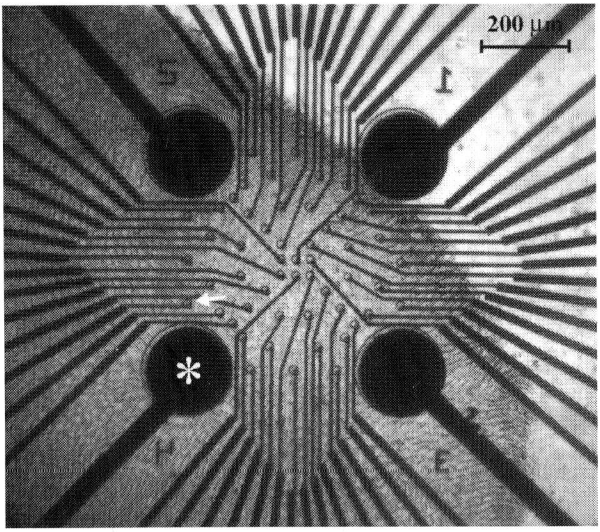

Figure 8. Isolated larval tiger salamander retina (darker region at left) overlying parylene-based platinum electrode array. Arrow indicates 10 μm diameter electrode used for recording trace in Figure 9. Asterisk identifies 200 μm diameter stimulating electrode used to generate action potentials seen inFigure 9.

With the lights off, a 20 μA, 400 μs/phase, cathodic-first biphasic electrical pulse was applied between the stimulating electrode indicated with an asterisk in Figure 8 and the ground electrode. The voltage trace from the recording electrode is shown in Figure 9. This stimulation was consistently repeatable over a 50 pulse train with a 400 ms inter-pulse interval, and other stimulating electrodes were also capable of "epiretinally" stimulating other cells in the retinal slice. As is clear from these results, the parylene-based platinum electrode was able to stimulate the tissue and elicit a response similar to the response generated from a light pulse in this intact retina. Given these results and the knowledge garnered from clinical trials with prototype arrays fabricated of other materials, it is not unreasonable to presume that our arrays will most likely be able to stimulate retinal tissue in other species, including human.

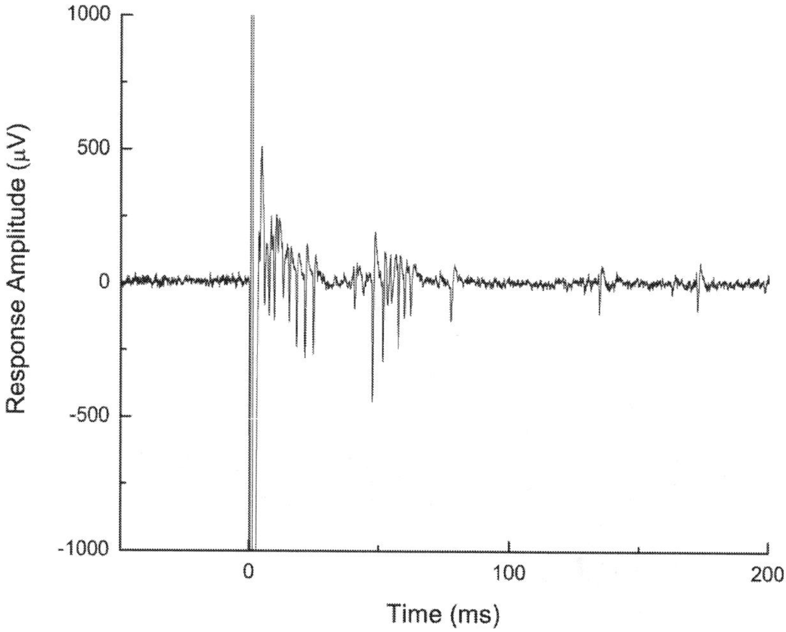

Figure 9. Typical recording of response of cells overlying recording electrode to a 20 μA, 400 μs/phase, cathodic-first biphasic electrical pulse from "epiretinal" stimulating parylene-based platinum electrode denoted with an asterisk in Figure 8.

SPINAL CORD PROSTHESIS

System Overview

The ideal spinal cord stimulation system, just like the retinal system, would have a power source, circuitry for driving the appropriate electrodes, as well as a cable and electrode array, this time implanted epidurally or subdurally on the spinal cord. We believe that a penetrating electrode array would be problematic for implantation and would likely lose efficacy and fail ultimately due to a gliosis over time, as has been shown in many other studies [27]. The power source could be an RF coil, or could, due to the much larger space available in the abdomen and back as compared with the eye, be a rechargeable battery capable of charging through the inductive link. The RF coil, in addition, would enable reprogramming of the implanted electronics for alternative stimulation protocols at the physician's discretion. The electrode array should be conformable to the spinal cord so that it can tonically stimulate at low currents and with high precision. While a completely implantable system is the ultimate goal, an interim goal is to stimulate the spinal cord chronically from an array connected to a head plug, while simultaneously being able to record electromyograms (EMGs). In order to achieve this, we have studied the efficacy of the multielectrode array portion of this system and have begun to develop a connector technology capable of connecting 36 electrodes in with a small enough form factor to be chronically mounted on a mouse skull.

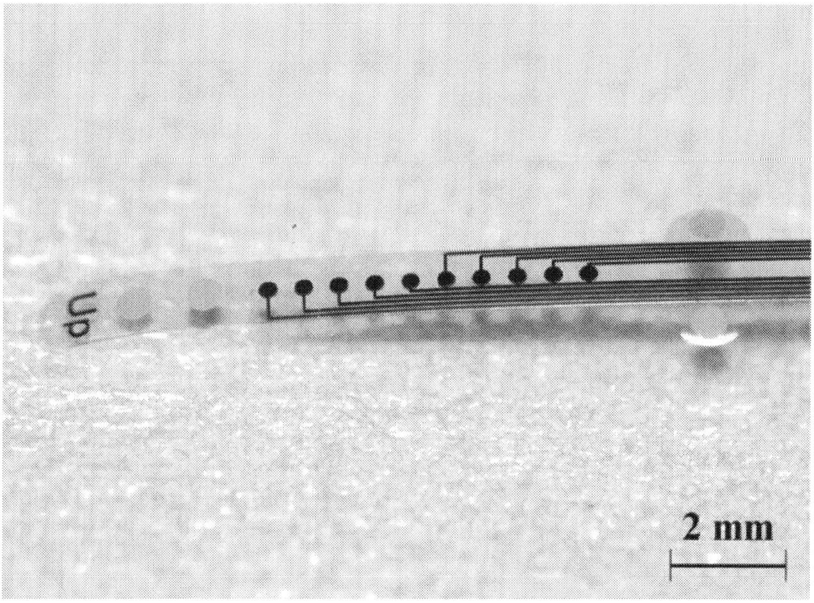

Figure 10. Parylene MEA for murine spinal cord stimulation and recording.

Fabrication

Spinal cord arrays, consisting of five or ten electrodes of 250 μm diameter were designed and fabricated (Figure 10). Interelectrode spacing was controlled so that each array of electrodes covered four to five segments of the murine lumbosacral spinal cord upon implantation. Suture holes were also designed into the body of the array to ensure placement and attachment of the array on the cord, as well as to facilitate implantation (suture can be attached to the end of the array and can be threaded along the cord first to help direct the array along it).

The single-metal layer fabrication process was performed using a contact aligner process for fast throughput. The fabricated arrays were annealed to increase the adhesion of parylene to parylene. At the same time, they were clamped between two pieces of Teflon or glass slides coated with aluminum foil to ensure they would be flat during implantation. The arrays were connected via Clincher connectors (FCI, Versailles Cedex, France) to the stimulation and recording electronics.

Implantation and Testing

Just prior to implantation, the arrays were rinsed in isopropyl alcohol. Under isoflurane anesthesia, the spinal cord electrode arrays were implanted epidurally on spinal cord segments L2-S1 in nontransected mice. The electrodes were oriented linearly along the rostrocaudal extent of the cord. Recording capability was assessed by using the electrode array to record spinal cord potentials evoked by tibial nerve stimulation. Following stimulation of the tibial nerve, somatosensory evoked potentials were recorded from the cord dorsum at three lumbosacral levels (P1-P3, rostral to caudal). The recorded waveform consisted of three response peaks, two of which are clearly depicted in Figure 11 (N1 and N3). These findings closely mirror results reported previously in a study using conventional spinal cord recording electrodes [46] demonstrating that the recording capability of the array electrodes matches that of conventional electrodes. By measuring the difference in the response latencies obtained at each electrode position (corresponding to different levels of the spinal cord), and by utilizing the known, fixed interelectrode spacing, accurate measurements of the conduction velocities were obtained. The properties of these responses can potentially be used to diagnose the progressive recovery of the spinal cord as a result of treatments provided after a spinal cord injury.

To test the capability of the electrode array to act as a multichannel stimulating device for generating hindlimb movements, constant-current monophasic stimulus pulses (amplitude: 50-850 μA, frequency: 0.3-10 Hz, pulse duration: 0.5 ms) were applied to the spinal cord between each of the array electrodes and a ground electrode located near the shoulder, while muscle activity was monitored using electromyogram (EMG) recordings of the tibialis anterior and medial gastrocnemius muscles. Stimulation generated a typical three-component EMG action potential consisting of an early (direct motor), a middle (monosynaptic), and a late (polysynaptic) response, classified by post-stimulus latency (Figure 12). These data clearly indicate that the parylene arrays were able to stimulate the spinal cord in such a way that the musculature was activated.

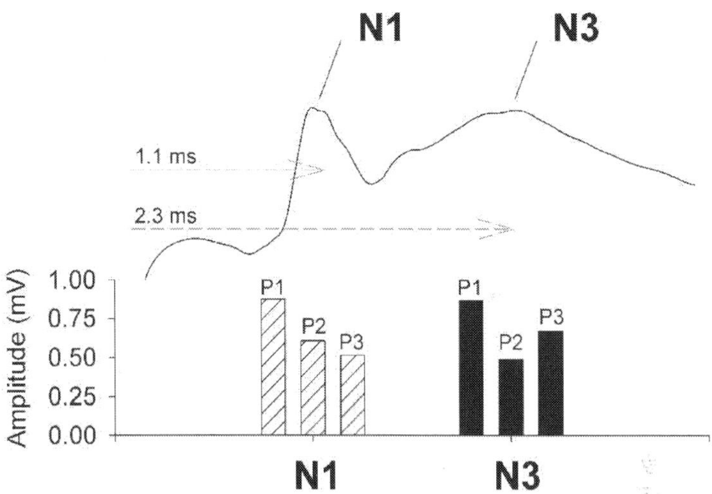

Figure 11. Peak amplitudes of somatosensory evoked potentials (N1 and N3) recorded from three levels of the rostrocaudal spinal cord (P1-P3). Example waveform at top shows approximate response times.

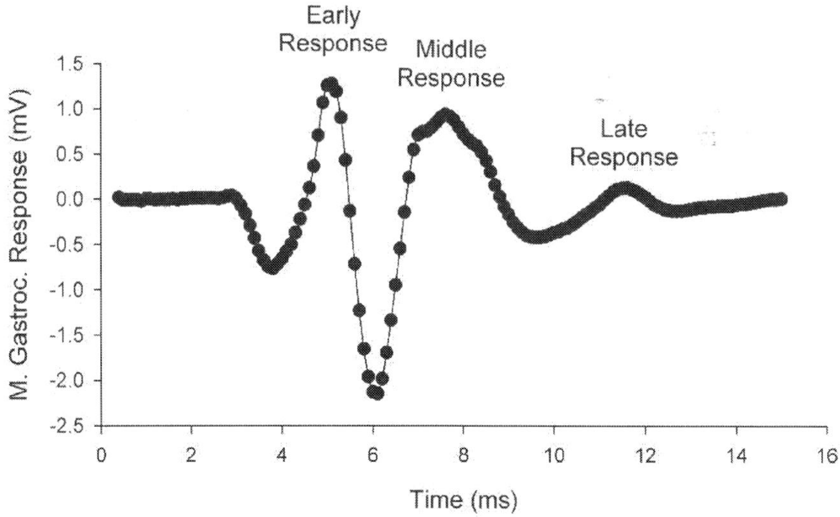

Figure 12. Typical medial gastrocnemius (ankle plantarflexor) EMG recording showing early, middle, and late responses after stimulation of spinal cord with parylene MEA.

Because of the known spacing of the electrodes on the array (as compared with traditional fine-wire electrodes which do not have known interelectrode spacing), we were able, in addition, to determine whether electrode position had a significant impact on muscle recruitment. The appearance and magnitude of each of the EMG responses was indeed correlated with the choice of electrode position (Figure 13). This serves as evidence that position of stimulation is very important. With a one-dimensional array, it is difficult to assess whether a bilateral stimulation paradigm would also result in lateralization of response, but we strongly suspect that this would be the case.

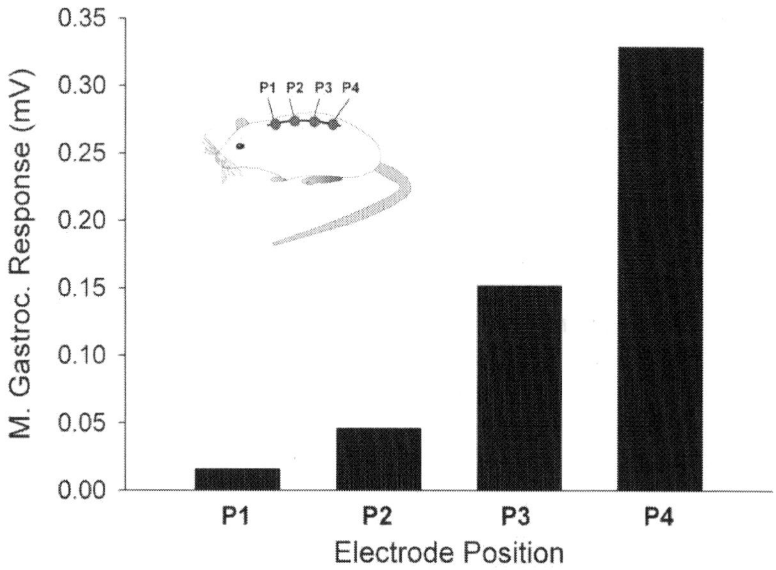

Figure 13. Medial gastrocnemius EMG showing varying levels of activation due to stimulation at different rostrocaudally located electrode sites.

DUAL-METAL-LAYER FLEXIBLE MEAS

Motivation

An inherent problem with single-metal-layer arrays is that electrodes and traces are necessarily fabricated alongside each other. For high-lead-count devices, this limits the size of electrodes and tends to crowd electrodes and traces into artificial groups. To allay these problems, a dual-metal-layer approach was devised that enables traces to pass underneath overlying electrodes.

Fabrication

Dual-metal-layer electrode arrays are fabricated as shown in Figure 14. Approximately 8 μm of parylene C is first deposited on a silicon wafer with the optional photoresist sacrificial layer, forming the underside of the electrode array. A platinum or titanium-platinum metal liftoff process is used to define traces with 16 μm pitch and 2000 Å to 3000 Å thickness. A second parylene deposition (~1 μm) forms the insulation between the two metal layers. At this point, 6 μm by 6 μm vias are patterned in the insulation layer over the ends of the traces using an O_2 plasma RIE. A second step-coverage optimized liftoff process is used to define a second metal layer comprising electrodes and traces, while at the same time achieving electrical continuity between the underlying traces and the overlying electrodes. A final parylene coating approximately 7 μm thick forms the top insulation. The electrodes are exposed and the overall geometry of the implant is defined in a final set of O_2 reactive-ion etches using a thick photoresist etch mask. Finally, the arrays are peeled from the wafer in a water bath or released through removal of the sacrificial photoresist in acetone. The process depends on optimal step coverage of the parylene sidewall during evaporation, which is aided, in part, by the slightly isotropic nature of the O_2plasma etch of parylene [47] as well as by the special design of the rotating wafer domes inside the e-beam evaporator, for which the angle of attack of the metal evaporant is adjusted for best coverage. This requirement conflicts marginally with those for successful metal liftoff, however, in this case, the liftoff technique is robust even under these step-coverage optimized conditions due to the choice of an LOR/positive photoresist compound layer.

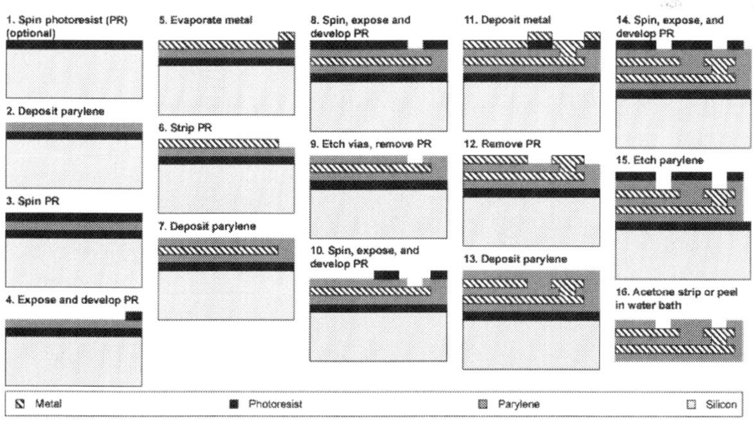

Figure 14. Fabrication process for parylene-based dual-metal-layer flexible MEAs.

Implantation and Testing

Chronically implantable retinal electrode arrays comprising 1024 75 μm diameter electrodes arranged in a complex biomimetic pattern that closely mimics the density of RGC in the human retina [48] were designed (the electrode density varied radially in a ratio matched to that of the RGCs), as shown at the left in Figure 15. These arrays (shown at right in Figure 15) were fabricated according to the dual-layer process, with 60 of the electrodes connected via two traces each to facilitate electrical conductivity verification. The strength of metal adhesion was verified using a Scotch tape test, which demonstrated that direct platinum evaporation is feasible without the need for a titanium adhesion layer. Electrical testing demonstrated a typical line impedance of a contact-electrode-contact circuit to be approximately 5 kΩ, which included two 8 μm wide traces of 20 mm length, as well as two via step junctions connecting underlying traces to the overlying electrode. Two types of via and electrode configurations were tested. Some electrodes (electrode SEM given in Figure 16 (left)) had vias connecting to the underlying trace near the center of the electrode, hence enabling charge spreading from the center of the electrode. One possible drawback to this configuration is that the contact from trace to electrode over the sidewall is a potentially vulnerable point of the circuit during processing (e.g., subsequent RIE processes) and during pulsing in electrolyte because the metal may be thinner there. The other electrode configuration had vias located adjacent to the electrode (electrode SEM given in Figure 16 (right)), with the possible advantage that it would be protected during RIE and subsequent pulsing by the overlying conformal parylene layer. An SEM showing the morphology of a single central via is given in Figure 17. This clearly depicts the sidewall coverage responsible for the electrical contact between the first and the second metal layers. In both configurations, each via had an impedance of less than 12.5 Ω. The best final configuration has not yet been determined.

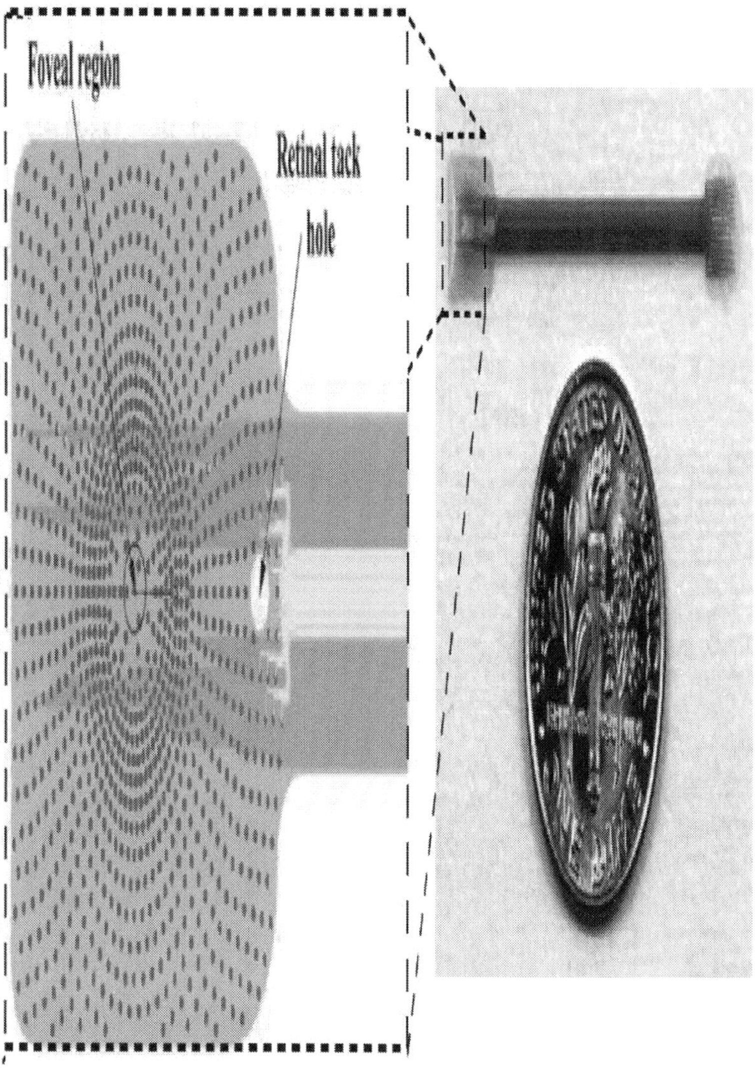

Figure 15. (left) Design of biomimetic dual-metal-layer retinal electrode array showing biomimetic arrangement of electrodes. (right) Fabricated biomimetic chronically implantable arrays with 60 of 1024 75 μm diameter electrodes connected through dual-layer process with U.S. dime for size comparison.

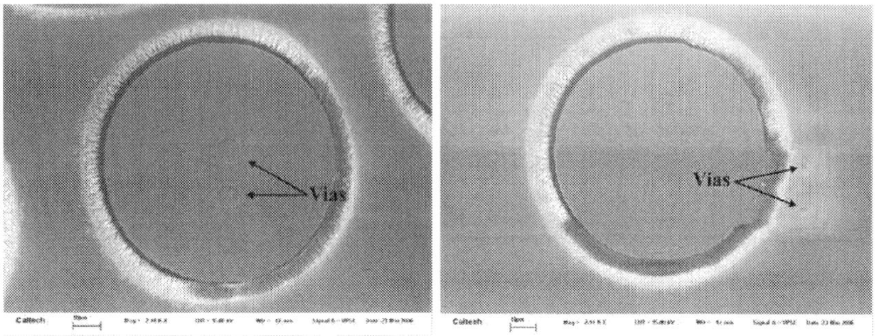

Figure 16. Two possible dual-layer electrode configurations. Electrode with central vias (left), and electrode with abutting vias (right).

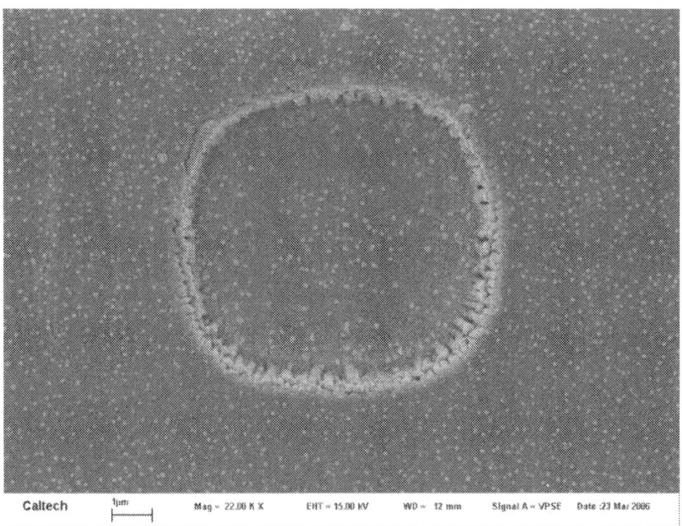

Figure 17. Magnified view of trace to electrode via showing sidewall coverage.

The arrays were successfully molded to the approximate curvature of the canine retina (Figure 18 (top)) using heat-annealing and a custom mold in a vacuum chamber, and sterilized using ethylene oxide gas. Two biomimetic arrays were implanted in the right eye of two canines through a 5 mm pars plana incision after vitrectomy, and were affixed to the retina (Figure 18 (bottom)) using a retinal tack modified by the addition of a PDMS washer (to account for the thin nature of the parylene arrays).

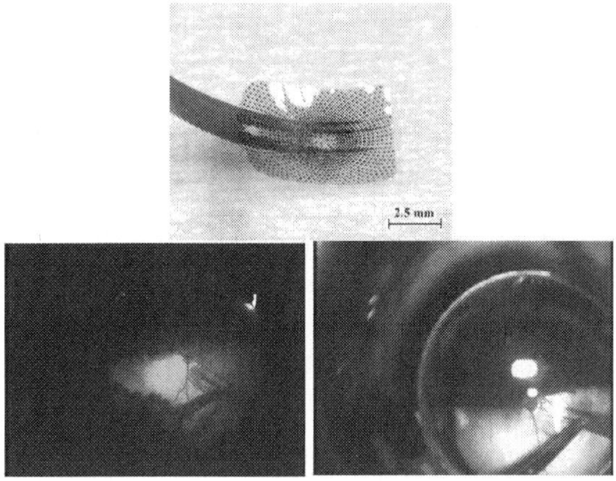

Figure 18. Heat-molded and annealed retinal electrode array with retained spherical curvature (top), and intraoperative photographs of tacking in each canine (bottom).

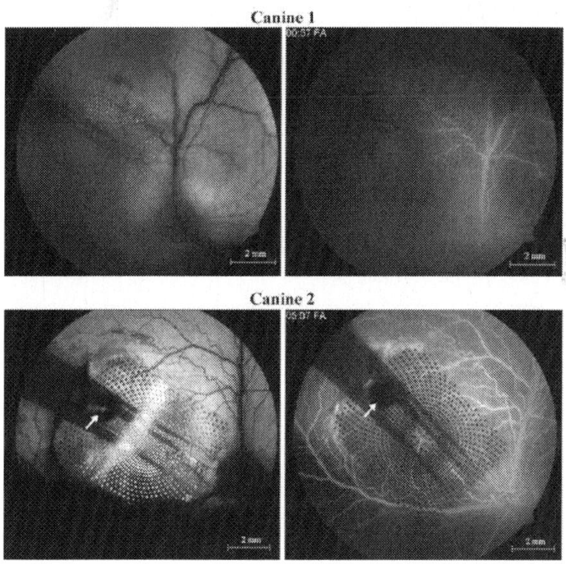

Figure 19. Fundus photographs (left) showing parylene MEAs tacked to the right retina of both animals and FAs (right) showing normal vessel perfusion under the arrays. Arrows point to retinal tacks.

Follow-up in both animals was conducted for six months using fundus photography, fluorescein angiography (FA), in which blood is fluorescently stained to assess vessel perfusion in the retina, and optical coherence tomography (OCT), an interferometric technique that enables cross-sectional imaging of the retina. Fundus photography and FAs of both animals, examples of which are shown in Figure 19, consistently demonstrated that vessel filling underneath the array was normal. Obstruction and vessel leakage would have been visualized if the array were placing excessive pressure on the retina. In addition, OCT demonstrated that the electrodes were consistently less than 50 μm away from the ganglion cell layer in both animals (typical OCTs of both animals are shown in Figure 20), an outcome that theoretically would afford excellent electrical coupling between the electrodes and the electrically excitable cells of the retina. It is important to note that in the OCT of the second canine, the scan was taken along a segment furthest from the tack site, where one might expect the least proximity. Even at this location, this array remained in very close apposition throughout the six-month implantation. Post-enucleation histology has since confirmed the excellent biostability seen during follow-up.

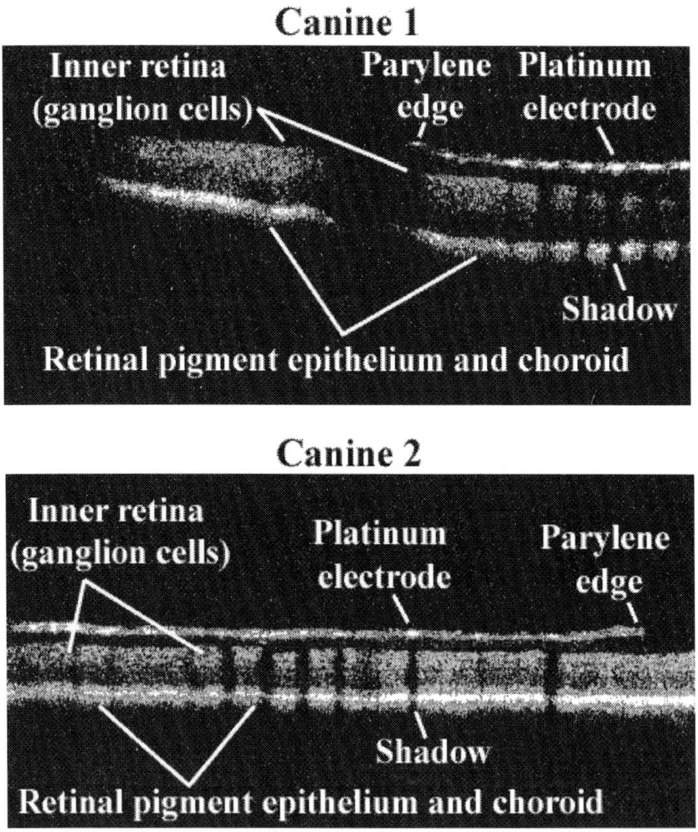

Figure 20. OCTs of both animals showing very close apposition (<50 μm) of the arrays to the RGC layer.

As was briefly mentioned previously, the dual-layer process proffers considerable advantages over the more traditional single-layer approach. Design of single-layer electrode arrays is usually hindered by the need to route traces amongst the electrodes. This tends to cause crowding of traces and electrodes into groups, an organization that may not be optimal for stimulating the tissue of interest. In addition, this has a propensity to constrain the geometric area of the electrodes in the MEAs to smaller sizes, and thus reduces the number of electrodes possible in a given area. The dual-layer process obviates these problems by enabling traces to pass under overlying electrodes without making contact to them, having the effect of both relaxing the constraints on electrode size and number and enabling more complex electrode organization (such as the biomimetic one presented in this work). Although the arrays fabricated here had just 60 electrodes of connectivity with 120 traces total, this was without making full use of both layers for wire routing and connection of electrodes. In order to not make traces unnecessarily narrow and of too high impedance, we believe an extension of this process to three or more metal layers will be necessary to achieve 1024 electrodes of total connectivity. Indeed, this fabrication process is easily extendable to create such structures through addition of extra layers of parylene and metal. Given the encouraging biostability results presented here and the ability of these arrays to stimulate retinal tissue, future studies will include chronic stimulation from implanted parylene-based arrays in an animal model.

Electroplating

As a possible mechanism for extending the longevity of chronically pulsed electrodes, we have investigated electroplated films of high surface-area platinum. Specially designed thin-film platinum electrode arrays, consisting of sixteen 75 μm and 150 μm diameter electrodes of 3000 μm center-to-center spacing, were fabricated according to the single-layer process. Initial experiments were performed on these arrays to determine material morphologies after plating at different potentials in an aqueous ammonium hexachloroplatinate solution according to Whalen *et al.* [49]. Subsequently, arrays were immersed in the solution in a specialized jig and six were plated at a plating potential of -0.6 V (vs. an Ag/AgCl reference electrode) for 1.5 hours. The others remained unplated. Electrochemical tests were performed to evaluate the efficacy of this plating step in extending electrode longevity under chronic pulsing.

The electrode morphologies of a typical array of 16 electrodes plated at different potentials are shown inFigure 21. Note that the 4 corner electrodes (1, 4, 13, and 16) were not plated. Magnified views of some of the possible morphologies attainable via this mechanism are shown in Figure 22. These micrographs show morphologies that likely correspond to a drastically increased surface area. In order to confirm this cyclic voltammograms (CVs) in O_2-free H_2SO_4 were taken of the electrodes before and after platinization. According to [50], "real" electrode surface area can be adequately estimated by integration of the area over the hydrogen adsorption peak or likewise the area under the hydrogen desorption peak. The two CVs in Figure 23 show that there was a more than 40-fold increase in surface area after platinization when compared with the pre-plated surface area (note the change in scale of the ordinate from nA to μA). Under pulsing, the voltage responses of both plated and unplated electrodes remained stable for approximately 29 days, at which point the unplated electrodes showed signs of failure. Voltage responses for one such electrode on day 26, 29, and 31 are overlaid in Figure 24 (left), which documents the progression of failure. The plated electrodes, on the other hand, remained intact for much longer, most surviving more than 50 days, or 430 million pulses, at which point the testing goal was met and the test was stopped. Overlaid voltage responses for one such electrode, showing the voltage responses at day 26, day 31, and day 50, are shown in Figure 25 (right). The electrochemical impedances at 1 kHz of a typical plated and unplated electrode are shown in Figure 25. A dramatic jump in impedance was observed for the unplated electrode at the time of failure, while the plated electrode demonstrated only minor variability in its lower impedance throughout the 430-million-pulse trial (most variability happened on the days that CVs were taken, as expected). These preliminary data corroborate the evidence that plating of the electrodes is beneficial to longevity, and suggest that high surface-area platinization of electrodes can have a dramatic effect on extending electrode life while lowering electrochemical impedance to charge delivery. Future work will include replication of these tests and chronic pulsing at high temperatures for longer times to further accelerate and assess the possible modes of failure.

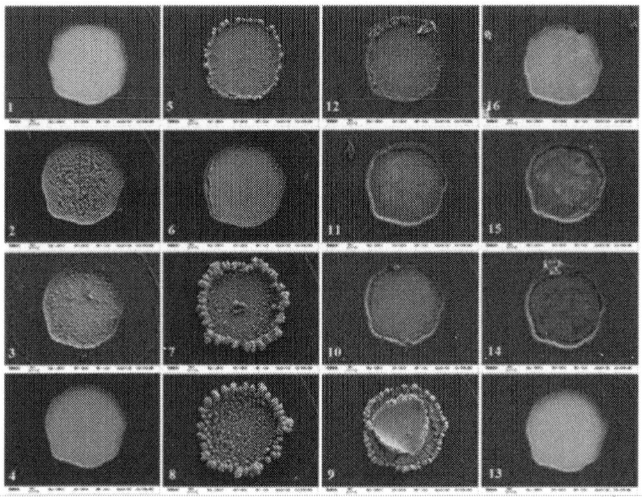

Figure 21. SEM of each of the electrodes on a typical 16-electrode array after platinization.

Figure 22. Magnified views of possible surface morphologies after platinization.

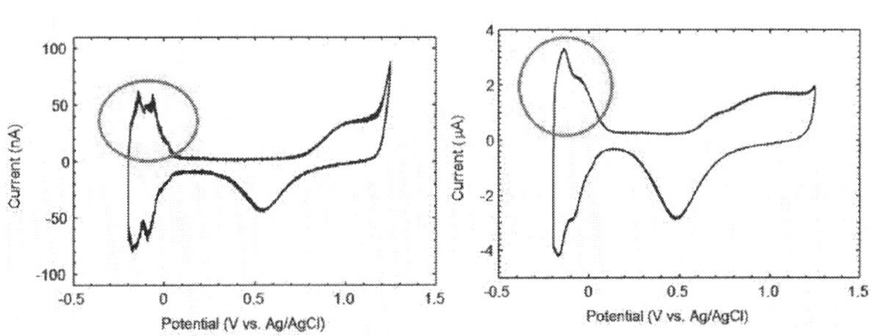

Figure 23. CVs showing more than 40-fold increase in electrode surface area from before plating (left) and after plating (right). The surface area is estimated by integrating the area under the peaks circled in red. Note change in scale of the ordinate. Scan rate: 100 mV/s. Electrolyte: O_2-free H_2SO_4 (N_2-bubbled).

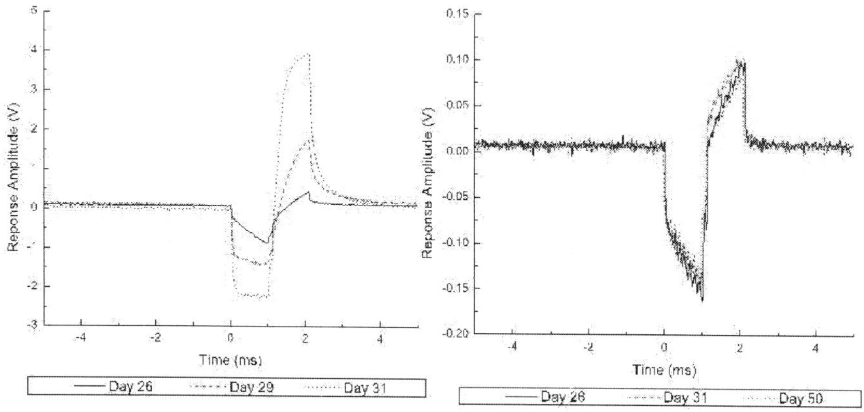

Figure 24. Voltage responses to a current pulse for (left) an unplated electrode, documenting the process of electrode failure, and (right) a plated electrode, showing steady responses throughout the 50 day test. Note response amplitudes for plated electrodes are far lower than those for the unplated electrode, as expected [27].

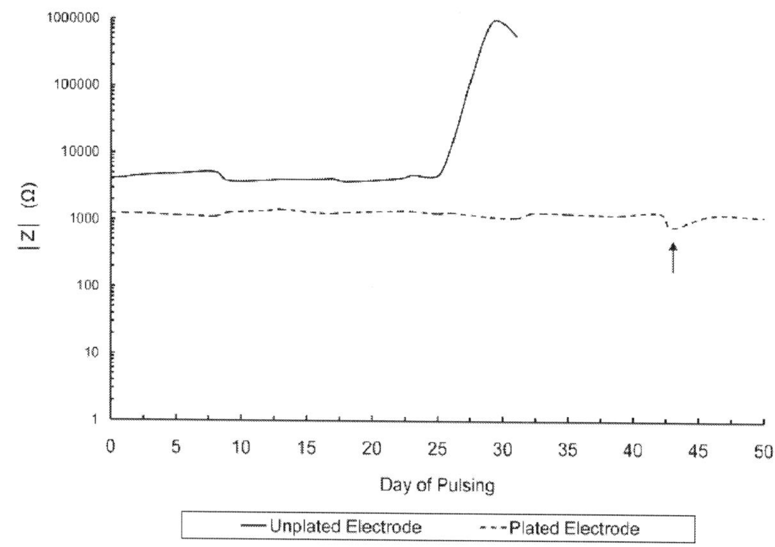

Figure 25. Magnitude of the electrochemical impedances at 1 kHz of an unplated and plated electrode over time. The unplated electrode showed a dramatic increase in impedance around day 30, at which time the test was stopped, whereas the plated electrode showed steady impedance through day 50. The arrow denotes a temporary dip in impedance due to CV scanning.

CHIP-LEVEL INTEGRATED INTERCONNECT TECHNOLOGY

Introduction

Despite our ability to fabricate such a large number of electrodes in such a small area, a significant impediment to future progress is the problem of how to package and interconnect these multielectrode arrays with foundry-fabricated ASICs, discrete components (e.g., chip capacitors, oscillators, diodes) and RF coils in a way that provides for high lead-count interconnects. A wafer-level process is cost prohibitive, as it is necessary to maximize the area of a wafer devoted to IC processing to keep costs low. Furthermore, current technologies for packaging would be far too tedious and low yield to apply to a 1000-electrode device. In order to achieve our goal of a 1000-electrode retinal prosthesis and a high-density spinal cord stimulation system, then, a new way of forming such a package so as to enable high-lead-count integration is necessary.

We have invented a way to place prefabricated chips, manufactured, for example, at a foundry, into the fabrication process of a parylene-based multielectrode array and/or RF coil, such that all interconnections to the chip are made using standard photolithography and standard microfabrication techniques in a fully scalable manner [40]. This packaging scheme is known as the chip-level integrated interconnect (CL-I^2) package. Figure 26 shows an overview of the fabrication process and how multiple chips could be joined together in this manner. A detailed discussion of the fabrication process, as adapted from [40], follows.

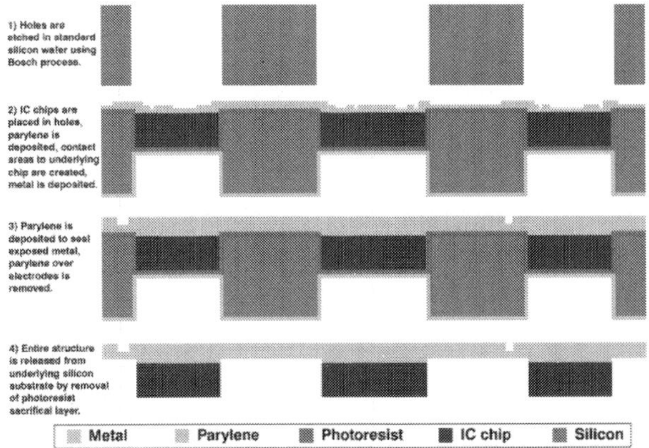

Figure 26. Overview of the CL-I^2 process. Multiple chip connections are possible.

Fabrication

Three MOSIS-fabricated ASICs, as well as seven chips fabricated to simulate them (with circuitry that facilitated testing), were used to demonstrate the CL-I^2 packaging technology. In order to fabricate the replicas of the MOSIS chips, these chips were imaged using a WYKO interferometer (Veeco Instruments Inc., Woodbury, NY, USA), and were found to have mean dimensions of 2.500 mm in length, 2.617 mm in width, and 254.2 μm in total thickness.

One hundred angstroms of chrome and 2000 Å of gold were e-beam evaporated on a 260 μm thick silicon wafer. Using a photoresist mask, the metal was wet etched to pattern pads of the same size and in the same locations as on the MOSIS-fabricated chips (approximately 70–100 × 100 μm^2 with a center-to-center pad spacing of approximately 200 μm), as well as a pattern of short circuits connecting these pads to nearby pads. After stripping the photoresist, a second photoresist layer was spun on the wafer and patterned as a mask for a Bosch through-wafer etch in a PlasmaTherm SLR-770B deep reactive ion etching (DRIE) system (Unaxis Corporation, St. Petersburg, FL, USA). This etch defined the length, width, and thickness of the simulated chips as 2.49 mm, 2.61 mm, and 260 μm, respectively. Finally, the photoresist mask was removed from the individual chips. In this manner, chips comprising simple electrical shorts and intrinsic resistors were fabricated as our primary CL-I^2 package test structures (Figure 27.)

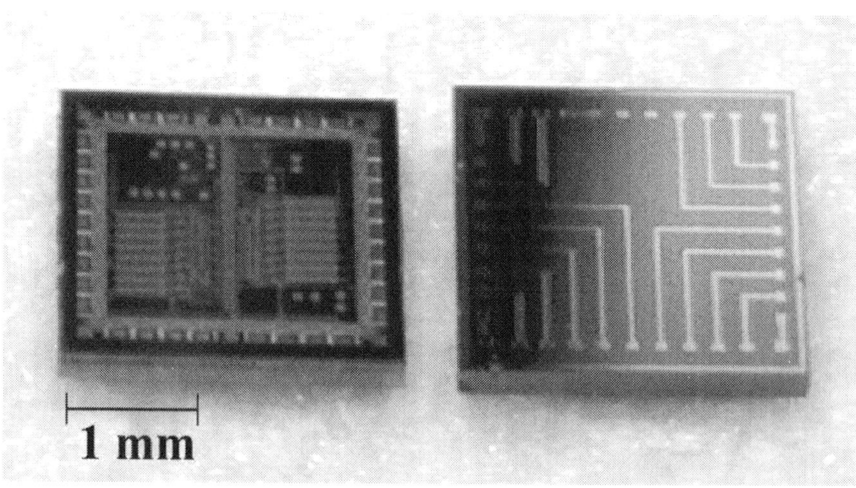

Figure 27. MOSIS ASIC (left) next to test chip (right).

The only properties of these prefabricated chips that had to be known before incorporation in the CL-I^{2}process were their overall length, width, and thickness, and the dimensions and locations of the contact pads. Figure 28 gives a detailed CL-I^2 process flow. To begin, shallow alignment marks are etched into a standard 550 μm thick silicon wafer using a thin photoresist mask and an SF_6 plasma. 2.51 × 2.63 mm^2 holes are then patterned after alignment in a 10X reduction stepper in thick photoresist and an optional silicon dioxide mask. Through holes are then etched using the Bosch DRIE process. After photoresist and oxide removal, Nitto tape is placed on the frontside of the wafer. The chips are then self-aligned in the holes by inserting them from the backside (the Nitto tape enables frontside planarization whereas the lateral dimensions of the etched cavity determine lateral displacement), and they are sealed in place using several drops of sacrificial photoresist to cover the backside of the chip and to fill the gaps around it. A subsequent approximately 12 μm thick parylene C deposition in a PDS2010 mechanically anchors the chips in place from the backside. After removal of the frontside parylene by peeling off the Nitto tape, vertical displacements of the chips are measured using a stylus profilometer (Alphastep 200 and P-15, KLA-Tencor, San Jose, CA, USA).

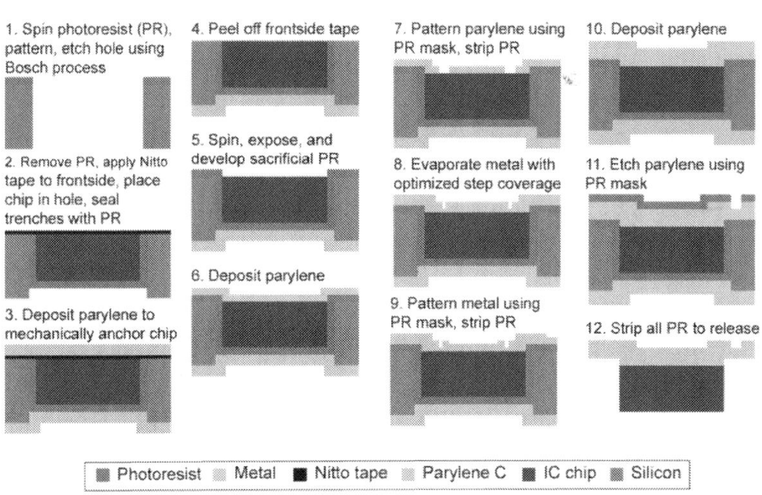

Figure 28. Detailed process flow for CL-I^2 package fabrication.

The parylene-based flexible electrodes, or, in this implementation, contact pads for electrical testing, are then fabricated on this wafer as if it were a whole wafer with prefabricated integrated circuitry. First, a photoresist sacrificial layer is spun on the wafer and patterned to expose the chip's surface. After baking to remove excess solvent, approximately 3 μm of parylene C is deposited on the entire wafer. Photoresist is spun on the wafer, exposed in the 10X reduction stepper, and developed to pattern etch holes above the on-chip pads, similar to the vias in the dual metal-layer process. This pattern is transferred into the parylene using an O_2 plasma in an RIE system, exposing the metal of these on-chip pads. Two hundred angstroms of titanium and 2000 Å of gold are deposited in the e-beam evaporation system using optimized step coverage, and patterned (using a photoresist mask and wet etching) to define the remote contact pads and remote pad to on-chip pad interconnects. The top photoresist is stripped, and a second layer of approximately 10 μm of parylene C is deposited and patterned as before, but this time to open the remote pads/electrodes to enable electrical testing. Finally, all photoresist, including the sacrificial layer, is removed by soaking the wafer in acetone, releasing a flexible parylene skin with embedded interconnects to the packaged ASIC. The host wafer can be substituted in the process with a precisely machined substrate, and can be reused after this release step. It is also important to note that the ASIC or discrete component can be of any thickness, but generally the thickness should be less than that of the host wafer or machined substrate, and it can have parylene or any hermetic coating deposited on it *a priori*, provided that the chip contacts can be opened using microfabrication techniques before the interconnect metal is laid down and patterned (Figure 28 steps 8 and 9). Thus, this technology combines the best aspects of chip-level packaging, in which every surface of the prefabricated chip can be manipulated or coated beforehand, and wafer-level packaging, in that photolithography and microfabrication can be performed on the surface after such chip-level techniques.

Integration Testing Results

We successfully performed photolithography on ten prefabricated stand-alone chips using this paradigm: seven test chips (three conformally coated in parylene *a priori*) and three MOSIS-fabricated chips (one coated in parylene *a priori*). A MOSIS chip anchored in place in the host substrate is shown centrally in Figure 29 with the host wafer shown on the perimeter. As is expected, minimizing vertical displacement of the ASIC from the wafer surface is crucial for further photolithography steps. Figure 30gives typical surface profiles of all ten chips with respect to the surrounding host wafer, and indicates that for most chips, this vertical displacement was less than 5 μm after removal of the frontside Nitto tape. Photolithography on the somewhat anomalous test chips 2 and 7, however, was also successful. The accurate horizontal alignment of the perimeter interconnects to the embedded chips is shown inFigure 31, with Figure 32 giving a detailed micrograph of a single interconnect for both the test chip (a) and the MOSIS chip (b). By design, the chips should be self-aligned to within 10 μm of lateral displacement; some chips were aligned far better than this, however others were misaligned worse than this. With tighter tolerances on the cavity sidewalls, or with chip-alignment lithographic equipment, this alignment error could be improved. The embedded chip with remote contact pads is shown in Figure 33, and Figure 34 depicts the flexibility of this package. Functional contacts to the chips were verified as described in [40].

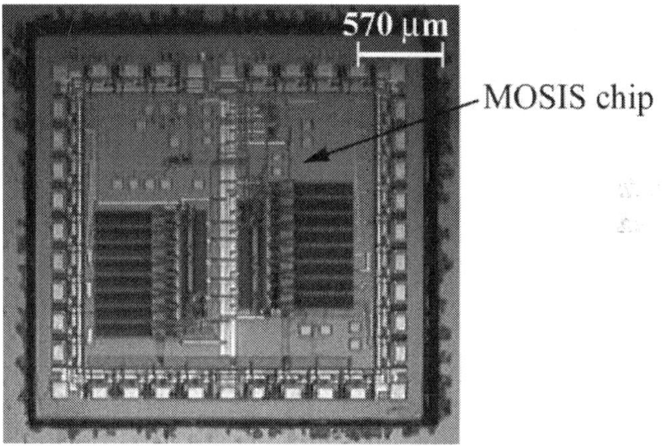

Figure 29. MOSIS chip (center) shown anchored in host silicon substrate (perimeter).

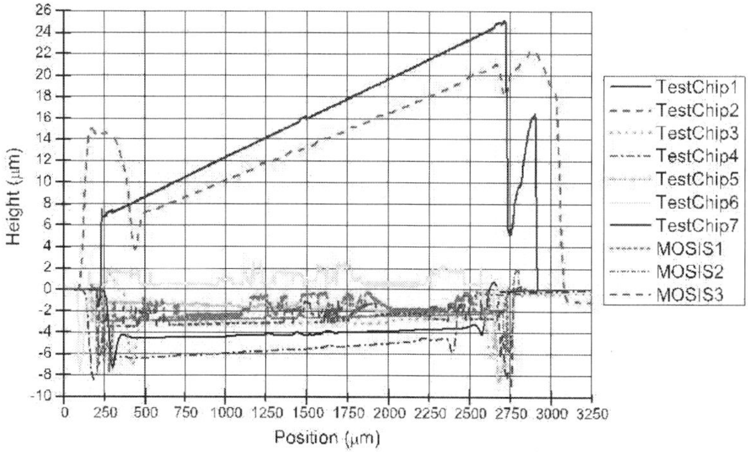

Figure 30. Typical single-axis vertical displacements of all 10 chips after mechanical anchoring in the host wafer (Figure 28, step 4), where the top surface of the wafer corresponds to 0 μm.

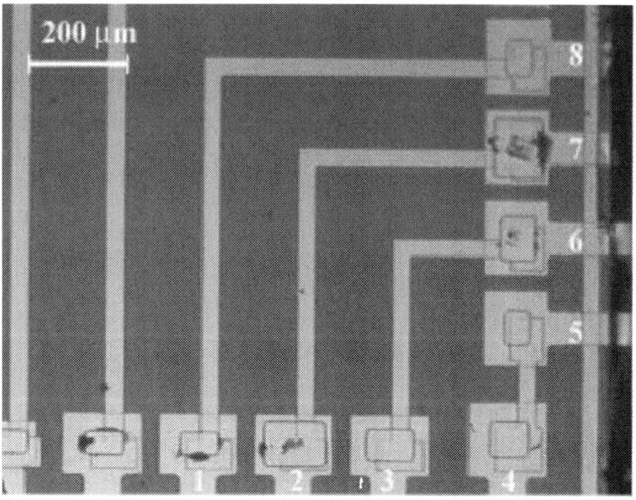

Figure 31. Embedded chip with fabricated perimeter interconnects (numbered traces connect to numbered remote pads shown in Figure 33).

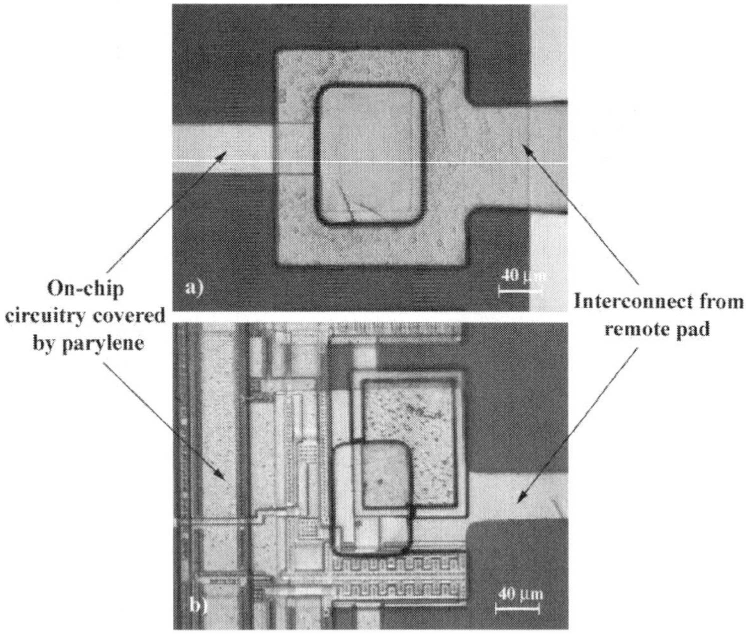

Figure 32. (a) Example of <10 µm lateral misalignment of a test chip; (b) Example of >10 µm lateral misalignment of a MOSIS chip.

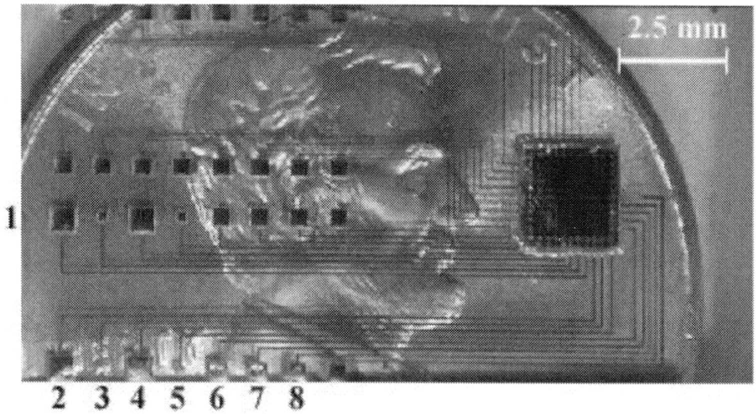

Figure 33. A CL-I^2 packaged chip shown resting on a penny.

Figure 34. Demonstrates flexibility of CL-I^2 package. Chip can be seen underlying overlying parylene "skin."

DISCUSSION

It should be stressed that the lead-count and interconnect density limitations for this technology stem only from the limitations of the microfabrication and photolithography equipment used to fabricate the CL-I^2 package, and, in particular, to pattern the first parylene etch (Figure 28, step 7). All interconnects to the chip are fabricated simultaneously during the metal deposition step, and depend on optimal step coverage of the parylene sidewall (aided in part by the slightly isotropic nature of the

O_2 plasma etch of parylene [47]). The CL-I^2 process thus avoids the use of tedious and comparatively low-density ball-wedge [38] or wire bonding.

Our method of incorporating discrete modules into a MEMS process is far more cost-effective when compared with full-wafer IC processing and MEMS integration [51], because valuable space on the wafer is not wasted during the IC fabrication step. Furthermore, in comparison to other ASIC integration attempts [51-53], this packaging scheme is superior for biodevices because it takes advantage of parylene's low water-absorption [54] and highly conformal pinhole-free deposition, and because the package is both flexible and biocompatible. Among the feasible uses for this technology is the interconnection of chips, devices such as other CMOS-compatible MEMS, as well as discrete components such as chip capacitors, fabricated using different materials and processes, to make large conglomerate circuits for neural prostheses and for other applications. This technology is capable of increasing the previously projected number of I/O interconnects available in 2010 significantly, while using lead-free, biocompatible materials. Fabrication is not limited to the use of parylene as either the backside anchoring material or as the frontside electrode insulation material, although, because of its superior electrical, mechanical, and water permeability properties when compared with other polymers, we believe parylene will ultimately prove to be the best choice for monolithic high-density neural prosthetics. It is interesting to note that another research group has, after our original publications [40,56, 57], explored an integration technique in polyimide very similar to ours, with interconnect density motivations much akin to our own [58]. We have recently demonstrated a fully integrated parylene-based single-channel neural stimulator [59]. *In vitro* measurements demonstrated the ability to generate 7 V pulses of 500 μs pulse width from a wireless transmitter 4 mm away. Preliminary evidence from animal implantation studies has shown these are mechanically reliable under surgical conditions.

Conclusions and Future Work

BioMEMS, an emerging field in which MEMS are designed, fabricated, and utilized to interface with and examine biological systems, is a discipline replete with incredible possibilities, but also one that is fraught with potential pitfalls. This is no more true than in the case of implantable microelectronics, where applications abound because of the near perfect match in the sizes of the functional components of the body, namely cells and neurons, and the technologies possible using microfabrication techniques. Materials must be carefully selected such that they are biocompatible with the body, while still enabling maximal functionality to be delivered to the patient, requirements that are often competing in nature. In this vein, the microtechnologies necessary for parylene-based flexible microelectrode technologies have been presented. All evidence thus far points to the fact that such parylene-based technologies are likely an ideal option for implantable neuroprostheses and microdevices.

It should be stressed that these technologies are not limited to use in retinal and spinal cord prostheses. In fact, the ability of our flexible arrays to conform to the geometries of interest in the human body enables them to be used in a variety of locations heretofore previously inaccessible with such high precision. Such locations include the surfaces of the cerebral cortex, another area of interest from both a scientific and treatment point of view due to such neurological problems as stroke, epilepsy, and memory loss. Areas of other interest include peripheral nerve and muscle. It is also possible to embed sensors in such arrays, as has recently been demonstrated [60], to assess mechanical forces placed on the tissues of interest by our arrays as well as to detect extrinsic pressures, such as those within the eye or within blood vessels.

Understandably, there is a public reticence to the implementation of such technologies in the human body. While such apprehension is not a recent phenomenon, the burgeoning era of computerized special effects in television and cinema has helped fuel the fear that the blending of "man" and "machine" can have devastating consequences. What is missed in such intimations is that, in the hands of ethical doctors, engineers, and other scientists, such consequences are extremely remote. But it is not about the inventors of these technologies, and it should never be about personal glory. All of that slips away the moment one talks to a person who has devoted their life to be a pioneer in the field by volunteering to be a test subject of such devices for the benefit of mankind. In such conversations, one realizes the full potential of this technology. Investigation into these devices not only has the possibility to positively affect the lives of such people, by enabling them to "walk" or "see" again, but it transcends all that by bringing about in all involved a sense of camaraderie. Indeed, the selfless motivation of such individuals who devote their most precious commodity, their body, to such studies, more than anything else highlights our very humanity. It shows that, despite the need for such technological innovation, it is only by working together and for one another that we, as a people, can break the bonds of human disease.

REFERENCES

1. The Eye Diseases Prevalence Research Group, "Prevalence of age-related macular degeneration in the United States," Arch Ophthalmol, vol. 122, pp. 564-572, April 1 2004.
2. G. Fishman, V. Vasquez, M. Fishman, and D. Berger, "Visual loss and foveal lesions in Usher's syndrome," Br J Ophthalmol, vol. 63, pp. 484-488, Jul 1 1979.
3. B. W. Jones, C. B. Watt, J. M. Frederick, W. Baehr, C.-K. Chen, E. M. Levine, A. H. Milam, M. M. Lavail, and R. E. Marc, "Retinal remodeling triggered by photoreceptor degenerations," The Journal of Comparative Neurology, vol. 464, pp. 1-16, 2003.
4. E. Margalit and S. R. Sadda, "Retinal and Optic Nerve Diseases," Artificial Organs, vol. 27, pp. 963-974, 2003.
5. Santos, M. S. Humayun, E. de Juan, Jr., R. J. Greenburg, M. J. Marsh, I. B. Klock, and A. H. Milam, "Preservation of the inner retina in retinitis pigmentosa. A morphometric analysis," Arch Ophthalmol, vol. 115, pp. 511-515, April 1, 1997 1997.
6. R. E. MacLaren, R. A. Pearson, A. MacNeil, R. H. Douglas, T. E. Salt, M. Akimoto, A. Swaroop, J. C. Sowden, and R. R. Ali, "Retinal repair by transplantation of photoreceptor precursors," Nature, vol. 444, pp. 203-207, 2006.
7. M. Humayun, R. Propst, E. de Juan, Jr., K. McCormick, and D. Hickingbotham, "Bipolar surface electrical stimulation of the vertebrate retina," Arch Ophthalmol, vol. 112, pp. 110-116, Jan 1 1994.
8. M. S. Humayun, E. de Juan, Jr., G. Dagnelie, R. J. Greenberg, R. H. Propst, and D. H. Phillips, "Visual perception elicited by electrical stimulation of retina in blind humans," Arch Ophthalmol, vol. 114, pp. 40-46, January 1, 1996 1996.
9. M. A. Schiefer and W. M. Grill, "Sites of Neuronal Excitation by Epiretinal Electrical Stimulation," IEEE Transactions on Neural Systems and Rehabilitation Engineering, vol. 14, pp. 5-13, 2006.
10. Y. Chow and V. Y. Chow, "Subretinal electrical stimulation of the rabbit retina," Neuroscience Letters, vol. 225, pp. 13-16, Mar 28 1997.
11. Veraart, M. C. Wanet-Defalque, B. Gerard, A. Vanlierde, and J. Delbeke, "Pattern recognition with the optic nerve visual prosthesis," Artificial Organs, vol. 27, pp. 996-1004, Nov 2003.
12. E. M. Schmidt, M. J. Bak, F. T. Hambrecht, C. V. Kufta, D. K. ORourke, and P. Vallabhanath, "Feasibility of a visual prosthesis for the blind based on intracortical microstimulation of the visual cortex," Brain, vol. 119, pp. 507-522, Apr 1996.

13. E. M. Maynard, "Visual prostheses," Annual Review of Biomedical Engineering, vol. 3, pp. 145-168, 2001.
14. J. S. Pezaris and R. C. Reid, "Demonstration of artificial visual percepts generated through thalamic microstimulation," PNAS, vol. 104, pp. 7670-7675, May 1, 2007 2007.
15. M. S. Humayun, J. D. Weiland, G. Y. Fujii, R. Greenberg, R. Williamson, J. Little, B. Mech, V. Cimmarusti, G. Van Boemel, and G. Dagnelie, "Visual perception in a blind subject with a chronic microelectronic retinal prosthesis," Vision Research, vol. 43, pp. 2573-2581, Nov. 2003.
16. M. S. Humayun, R. Freda, I. Fine, A. Roy, G. Fujii, R. J. Greenberg, J. Little, B. Mech, J. D. Weiland, and E. de Juan, "Implanted intraocular retinal prosthesis in six blind subjects," Investigative Ophthalmology & Visual Science, vol. 46, p. 1144, 2005.
17. B. S. Wilson, C. C. Finley, D. T. Lawson, R. D. Wolford, D. K. Eddington, and W. M. Rabinowitz, "Better speech recognition with cochlear implants," Nature, vol. 352, pp. 236-8, Jul 18 1991.
18. J. D. Weiland, W. Liu, and M. S. Humayun, "Retinal prosthesis," Annual Review of Biomedical Engineering, vol. 7, pp. 361-401, 2005.
19. C. J. Lee, J. A. Vroom, H. A. Fishman, and S. F. Bent, "Determination of human lens capsule permeability and its feasibility as a replacement for Bruch's membrane," Biomaterials, vol. 27, pp. 1670-8, Mar 2006.
20. L. J. Noble and J. R. Wrathall, "Correlative Analyses of Lesion Development and Functional Status after Graded Spinal-Cord Contusive Injuries in the Rat," Experimental Neurology, vol. 103, pp. 34-40, Jan 1989.
21. K. Kurohane, A. Tominaga, K. Sato, J. R. North, Y. Namba, and N. Oku, "Photodynamic therapy targeted to tumor-induced angiogenic vessels.," Cancer Lett, vol. 167, pp. 49-56, Jun 2001.
22. K. E. Krishnan L, Jewell WR, "Immediate effect of irradiation on microvasculature.," Int J Radiat Oncol Biol Phys, vol. 15, pp. 1147-1450, Jul 1988.
23. J. W. Little, J. F. Ditunno, S. A. Stiens, and R. M. Harris, "Incomplete spinal cord injury: Neuronal mechanisms of motor recovery and hyperreflexia," Archives of Physical Medicine and Rehabilitation, vol. 80, pp. 587-599, May 1999.
24. J. H. Burridge, P. N. Taylor, S. A. Hagan, D. E. Wood, and I. D. Swain, "The effects of common peroneal stimulation on the effort and speed of walking: a randomized controlled trial with chronic hemiplegic patients," Clinical Rehabilitation, vol. 11, pp. 201-210, Aug 1997.
25. J. Rozman, R. Acimovicjanezic, I. Tekavcic, M. Kljajic, and M. Trlep, "Implantable Stimulator for Selective Stimulation of the Common Peroneal Nerve - a Preliminary-Report," Journal of Medical Engineering & Technology, vol. 18, pp. 47-53, Mar-Apr 1994.

26. P. Taylor, J. Burridge, A. Dunkerley, D. Wood, J. Norton, C. Singleton, and I. Swain, "Clinical audit of 5 years provision of the Odstock dropped foot stimulator," Artificial Organs, vol. 23, pp. 440-442, May 1999.
27. K. W. Horch and G. S. Dhillon, Neuroprosthetics theory and practice. River Edge, N.J.: World Scientific, 2004.
28. R. Brissot, P. Gallien, M. P. Le Bot, A. Beaubras, D. Laisne, J. Beillot, and J. Dassonville, "Clinical experience with functional electrical stimulation-assisted gait with parastep in spinal cord-injured patients," Spine, vol. 25, pp. 501-508, Feb 15 2000.
29. D. Graupe and K. H. Kohn, "Functional neuromuscular stimulator for short-distance ambulation by certain thoracic-level spinal-cord-injured paraplegics," Surgical Neurology, vol. 50, pp. 202-207, Sep 1998.
30. E. Roederer, N. H. Goldberg, and M. J. Cohen, "Modification of retrograde degeneration in transected spinal axons of the lamprey by applied DC current," J. Neurosci., vol. 3, pp. 153-160, January 1, 1983 1983.
31. C. C. Stichel and H. W. Muller, "Experimental strategies to promote axonal regeneration after traumatic central nervous system injury," Progress in Neurobiology, vol. 56, pp. 119-148, 1998.
32. S. A. Maguire AM, Radiation retinopathy. In: Retina. St. Louis: Mosby-Year book., 1994.
33. M. Szwarc, "Some Remarks on the CH2=Benzene=CH2 Molecule," Discussions of the Faraday Society, vol. 2, pp. 46-49, 1947.
34. W. F. Gorham, "A New General Synthetic Method for Preparation of Linear Poly-P-Xylylenes," Journal of Polymer Science Part A-1: Polymer Chemistry, vol. 4, pp. 3027-3039, 1966.
35. L. Wolgemuth, "Crystal-Clear Coating Covers Components," Medical Design, vol. 6, pp. 48-51, December 1 2006.
36. "New SCS Parylene HT Conformal Coating Ideal For High-Temperature Applications." vol. 2007: Specialty Coating Systems.
37. "Reliable protection for advanced electronics." vol. 2007: Specialty Coating Systems.
38. T. Stieglitz, W. Haberer, C. Lau, and M. Goertz, "Development of an inductively coupled epiretinal vision prosthesis," in Int. IEEE Eng. in Med. and Biol. Soc. Meet., San Francisco, CA, USA, 2004, pp. 4178-4181 Vol.6.
39. C. Hoogerwerf and K. D. Wise, "A three-dimensional microelectrode array for chronic neural recording," IEEE Transactions on Biomedical Engineering, vol. 41, pp. 1136-1146, 1994.
40. D. C. Rodger, J. D. Weiland, M. S. Humayun, and Y.-C. Tai, "Scalable high lead-count parylene package for retinal prostheses," Sensors and Actuators B: Chemical, vol. 117, pp. 107-114, 2006.
41. W. Li, D. C. Rodger, E. Meng, J. D. Weiland, M. S. Humayun, and Y.-C. Tai, "Flexible Parylene Packaged Intraocular Coil for Retinal Prostheses," in Int.

IEEE-EMBS Microtech. in Med. and Bio. Meet., Okinawa, Japan, 2006, pp. 105-108.

42. P. J. Chen, D. C. Rodger, M. S. Humayun, and Y. C. Tai, "Unpowered spiral-tube parylene pressure sensor for intraocular pressure sensing," Sensors and Actuators a-Physical, vol. 127, pp. 276-282, Mar 13 2006.
43. P. J. Chen, D. C. Rodger, E. M. Meng, M. S. Humayun, and Y. C. Tai, "Surface-micromachined parylene dual valves for on-chip unpowered microflow regulation," Journal of Microelectromechanical Systems, vol. 16, pp. 223-231, Apr 2007.
44. E. Meng, P.-J. Chen, D. Rodger, Y.-C. Tai, and M. Humayun, "Implantable parylene MEMS for glaucoma therapy," in Int. IEEE-EMBS Microtech. in Med. and Bio. Meet., Oahu, HI, USA, 2005, pp. 116-119.
45. K. Ahuja, M. R. Behrend, M. Kuroda, M. S. Humayun, and J. D. Weiland, "Spatial Response Properties of Electrically Stimulated Retina," Invest. Ophthalmol. Vis. Sci., vol. 48, p. 4444, May 10 2007.
46. P. Chandran, K. Oda, H. Shibasaki, and M. Pisharodi, "Spinal somatosensory evoked potentials in mice and their developmental changes," Brain Dev, vol. 16, pp. 44-51, Jan-Feb 1994.
47. E. Meng and Y.-C. Tai, "Parylene etching techniques for microfluidics and bioMEMS," in IEEE Int. Conf. on MEMS, Miami, FL, USA, 2005, pp. 568-571.
48. A. Curcio and K. A. Allen, "Topography of Ganglion-Cells in Human Retina," Journal of Comparative Neurology, vol. 300, pp. 5-25, Oct 1990.
49. J. J. Whalen, J. D. Weiland, and P. C. Searson, "Electrochemical Deposition of Platinum from Aqueous Ammonium Hexachloroplatinate Solution," Journal of The Electrochemical Society, vol. 152, pp. C738-C743, 2005.
50. S. B. Brummer and M. J. Turner, "Electrical Stimulation with Pt Electrodes: I-A Method for Determination of "real" electrode Areas," IEEE Transactions on Biomedical Engineering, vol. 24, pp. 436-439, 1977.
51. Y. Xu, Y.-C. Tai, A. Huang, and C.-M. Ho, "IC-integrated flexible shear-stress sensor skin," Microelectromechanical Systems, Journal of, vol. 12, pp. 740-747, 2003.
52. J. T. Butler, V. M. Bright, and J. H. Comtois, "Advanced multichip module packaging of microelectromechanical systems," in Int. Conf. on Solid-State Sensors and Actuators, Chicago, IL, USA, 1997, pp. 261-264.
53. W. Daum, W. E. Burdick, and R. A. Fillion, "Overlay high-density interconnect: a chips-first multichip module technology," Computer, vol. 26, pp. 23-29, 1993.
54. J. J. Licari and L. A. Hughes, Handbook of polymer coatings for electronics : chemistry, technology, and applications, 2nd ed. Park Ridge, N.J., U.S.A.: Noyes Publications, 1990.

55. R. R. Tummala, Fundamentals of microsystems packaging. New York: McGraw-Hill, 2001.
56. C. Rodger and Y. C. Tai, "Microelectronic packaging for retinal prostheses," IEEE Engineering in Medicine and Biology Magazine, vol. 24, pp. 52-57, Sep-Oct 2005.
57. C. Rodger, J. D. Weiland, M. S. Humayun, and Y.-C. Tai, "Scalable flexible chip-level parylene package for high lead count retinal prostheses," in Int. Conf. Solid-State Sensors, Actuators and Microsystems, Seoul, Korea, 2005, pp. 1973-1976.
58. H. Sharifi, T. Y. Choi, and S. Mohammadi, "Self-aligned wafer-level integration technology with high-density interconnects and embedded passives," IEEE Transactions on Advanced Packaging, vol. 30, pp. 11-18, Feb 2007.
59. W. Li, D. C. Rodger, A. Pinto, E. Meng, J. D. Weiland, M. S. Humayun, and Y.-C. Tai, "Parylene-based integrated wireless single-channel neurostimulator," Sensors and Actuators A: Physical, vol. 166, pp. 193-200.
60. B. J. Kim, C. A. Gutierrez, G. A. Gerhardt, and E. Meng, "Parylene-based electrochemical-MEMS force sensor array for assessing neural probe insertion mechanics," in Micro Electro Mechanical Systems (MEMS), 2012 IEEE 25th International Conference on, pp. 124-127.

CITATION

Damien C. Rodger, Wen Li, James D. Weiland, Mark S. Humayun and Yu-Chong Tai (2013). Flexible Circuit Technologies for Biomedical Applications, Advances in Micro/Nano Electromechanical Systems and Fabrication Technologies, Assistant Professor Kenichi Takahata (Ed.), ISBN: 978-953-51-1085-9, InTech, DOI: 10.5772/55308.

Index